Do-It-Yourself

HOME
PROTECTION

A Common-Sense Guide

By RALPH TREVES

Popular Science Publishing Company

Harper & Row • New York, London

SOURCES OF MATERIALS LISTED

Manufacturers of various products are mentioned throughout the text. The full addresses of these manufacturers are listed separately in the back of the book. In some instances, retail dealers in different sections of the country are included as a convenience to readers who seek a source of supply. Prices, when stated, are those listed in current catalogues, and may vary considerably in different areas.

Library of Congress Catalog Card Number 74-167604

Designed by Jeff Fitschen

Manufactured in the United States of America

Contents

I BURGLARPROOFING YOUR HOME

1

How Safe
is Your Home?

It is frightening and heartbreaking to return home one evening to find that someone has broken in, ransacked the closets, upturned all the dresser drawers, and made off with your silverware and clothing, your coin collection, and your wife's jewels.

This, unfortunately, is not an unusual occurrence. Each year, more than a million American homes (yes, a *million!*) are broken into and looted. More than half of these burglaries took place during the daytime. The number of burglaries has more than doubled in the past decade, resulting in annual losses of an estimated billion dollars in stolen articles and property damage. And no cost can be set on serious injuries to residents from physical encounters with the prowlers.

While the city apartment dweller still is the most frequent victim of housebreaking, the crime wave has spread rapidly into the formerly safe and placid suburbs, small towns, and even rural areas. Every section of the country is affected. There just aren't any "good" neighborhoods any more!

Even the home of a police chief is not immune. In San Jose, California, thieves pried open a window of Chief Blackmore's home, ransacked all the rooms, and made off with his stereo system, television set, and some cash. It was the fourth house on the chief's block to be broken into in less than two months.

Financially damaging as burglaries may be, a more serious consequence is the persistent fear, often of panic proportions, that now obsesses so many families, so that even your children, just as your wife and yourself, are unable to sleep peacefully, constantly alert for strange sounds in the night, expecting to be pounced upon at any moment by an intruder. Others, particularly single people living alone, are afraid to come home in the evening, for fear of encountering a burglar inside.

You're on Your Own. If nothing like this has happened, you've been exceedingly lucky, or have taken adequate protective measures. Many families not so fortunate have been aroused to the need for better home security only after sad experience. But for those who have so far escaped this trouble, it's foolhardy to continue relying on common locks, archaic window latches, and limited police protection. Police patrols are indeed an essential line of defense, and they keep up an unrelenting drive on prowlers. The F.B.I. has stated, however, in its annual U.S. *Crime Reports,* that law enforcement agencies no longer can cope adequately with this crime of stealth, and in fact have been able in recent years to apprehend only about 20 per cent of the offenders.

A Change for the Worse. Not many years ago families could feel quite safe and relaxed in their homes without making a big thing about security measures. Only the most simple protective routine was relied upon, and sometimes none at all, with rarely any dire consequences. Nighttime lockup was perfunctory—usually just slipping the door bolt or turning the key. Homeowners seldom bothered to close their downstairs windows, especially if there were screens, which were considered a sufficient barrier. Doors were left open all day, and often in the evening too. When the family went out, the door key more often than not was handily slipped under the mat or over the door frame.

Locks in many cases were of easily jimmied spring-latch design. While the front door might have a secure cylinder lock, the fact that the bolt had to be turned separately with the key to lock it from the outside, meant that it was mostly neglected as too inconvenient. The rear kitchen and porch doors were usually fitted with the kind of mortise lock that is opened with a "skeleton" key. (All locks are described in chapter 5.) Cellar doors more often than not were held with just a barrel bolt or hook-and-eye fastener. And even this kind of rudimentary security was not always utilized—if someone in the family remembered at the movies that he or she had forgotten to lock the door, that was no cause for worry.

Things are Different Now. You don't have to study the crime statistics to realize the extent of the burglary menace now. However, knowing the facts should not only encourage you to act for your own protection, it should provide clues to the direction that your defense efforts must take to assure effective results.

Burglary is defined in different ways in different states, but it is defined by the Crime Statistics Bureau as unlawful entry, or attempt at such entry, to commit a felony or theft, even though no force is used to gain entrance.

A total of 2,169,300 burglaries were reported in the F.B.I. *Unified Crime Reports* for the year 1970, of which 58 per cent or more than 1,200,000 were residence burglaries. The incidence of burglaries rose by 142 per cent from 1960 to 1970, including an 11 per cent rise from 1969 to 1970.

Not All Thefts Reported. Startling as it is, this figure of 1,200,000 home burglaries may be far too conservative. The President's Commission on Law Enforcement and Administration has said that property crimes were more than triple the rate indicated by the F.B.I. reports. The Commission report, titled *The Challenge of Crime in a Free Society,* cited detailed surveys in several major cities which showed that the burglary rate for individual victims was more than 300 per cent greater than that reported by the F.B.I.

The chief explanation for this disparity was that many victims failed to report burglary losses because they felt that the police "could not do anything" to halt the thefts or to recover the stolen articles. Some of the reluctance to report burglaries probably was due to fear that insurance companies would cancel their policies, which in a sense tends to inflict added punishment on the helpless victim rather than on the perpetrator of the crime.

There is another statistical factor of importance: when an unlawful entry includes a violent attack upon an occupant, the offense is listed as a robbery rather than a burglary. Such encounters with a desperate, cornered prowler often result in serious injuries to the resident. Fortunately, such incidents are comparatively infrequent, with less than 32,000 attacks on residents occurring during the more than a million home burglaries. This figure covers only cities encompassing less than half the total national population; data on rural and small town incidents, which are not available, might well double it.

Daytime Burglaries Soaring. Over the period from 1960 to 1970, daytime burglaries of residences increased by an astounding and ominous 337 per cent. The rise in the last year alone of that period was 13 per cent, an increase far greater than that for nighttime thefts. The obvious inferences to be drawn are that homeowners have not yet applied adequate security routines to the family daytime practices, and that the type of criminal involved is steadily changing from that of a nighttime prowler to a younger, opportunity-seeking element that can operate more openly without attracting suspicious attention.

This latter group apparently consists mainly of teenagers and youths, and even young girls, who may be taken as friends of neighbors, or as legitimate delivery or service personnel. In any event, the young person is less likely to be challenged as he walks to the side door of a home, or loiters in an apartment corridor. And it takes him (or her) just a moment to enter if the door is held with a flimsy lock that just needs a push to crash it open, or a plastic window screen that can be sliced apart with a razor blade.

Perhaps a contributing element is that more women are joining the labor force, leaving the home or apartment unattended and easy prey during the daytime hours, a time when the noise of breaking window glass or splintering doors will less likely be noticed. Also, there is the understandable tendency to relax security during the day, allowing doors to remain unlocked, windows open, and alarm turned off, even when no one is at home.

The marauders seem to have caught onto this. Instead of prowling under cover of darkness at night, they are finding it safer and easier to operate under the cover of daytime noise, the diverted attention of neighbors, and relaxed security measures.

A Year 'round Crime. Burglars know no season. The peak month in 1967 was December, but in the following year the peak month switched to July, and in 1970 was back to December. Draw whatever conclusions you may from that!

Every Section Affected. No part of the country is spared the soaring burglary incidence. In 1969, California had the highest burglary rate, with 1,676 incidents per 100,000 population, Florida was second with 1,358, and third place was tied by both Hawaii and New York with 1,304 each. The Southern states had 27 per cent of the total burglary volume, and in 1970 experienced the sharpest increase, 14 per cent, in the burglary rate.

Suburbs, Small Towns Feel Rise. In every type of community, and every stratum of society, homes are no longer safe from intrusion. Big cities, those over 250,000 population, have always borne the brunt of this crime, both in number of incidents and rate proportionate to the population. The rural areas fare best, with an average of 434 per 100,000 population, compared to 872 for suburban centers, and 1,948 for metropolitan cities. In the most recent report, however, suburban and rural areas suffered the largest rate of increase, each up 12 per cent.

In the cities, housebreaking isn't limited to the congested tenement districts, New York City's Park Avenue apartments and expensive hotels are constant victims despite their extensive protective arrangements of doormen, elevator operators, maids and butlers. Recent incidents were the robberies of Sophia Loren and Zsa Zsa Gabor in famous Park Avenue hotels. One result has been the widespread installations of burglar alarms in the more affluent private home sections. Now the alarm bell boxes high on outside walls are almost as common as television aerials.

In California's movieland, the palatial homes of famed film personalities in Beverly Hills, Bel Air, Newport, and Brentwood, are broken into time and again. Recent targets were the homes of John Wayne, Gregory Peck and Joe E. Brown. One of the 1970 Hollywood burglaries resulted in the loss of a Raphael painting, "The Peruzzi Madonna," valued at $1.2 million dollars. The burglary occurred while the owner was away from home for just one hour. Quite a number of lesser starlets, living in modest Hollywood apartments, have found on returning from work that their belongings had been ransacked by an intruder. Their greatest fear is that of coming home while the burglar is still in the apartment, or awakening at night and surprising a dope-crazed prowler.

Crash Any Barrier? You often hear the remark that "There's no way to stop a burglar—if he wants to get in, he can always do it." This defeatist notion is not justified. Instead, a determined and intelligent effort usually can overcome the menace, because every householder has definite advantages on his side.

A burglar will tackle any obstacle if he believes there's plenty of loot to be had. So your first objective is to avoid letting information get around, in the neighborhood and elsewhere, about any large sums of money, valuable jewelry, or other marketable possessions. In fact, you should make every effort to avoid attracting interest in your home for a "caper," as the street denizens say.

And while thieves are capable of breaking through almost any barriers when they are determined to do so, it is also a fact that they tend to pick the easiest and least risky targets. So your second objective is to make your home as impregnable as you can, so that entry would be obviously difficult, noisy, and dangerous.

The practical defense measures you must take vary according to the particular circumstances in your home. Knowing the nature of the burglar provides a measure of his capabilities and methods to be counteracted. The house burglar of today is nearly always a delinquent teenager or drifting youth, poorly equipped, relying mainly on stealth and opportunity. But never forget that he doesn't hesitate to splinter a door or break a window glass to get inside.

This is not to imply that all burglaries are casual and easygoing, aimed at a quick grab for whatever loot can be obtained. On the contrary, many seem to be planned with accurate knowledge about a large sum of money or particular jewelry in the house, and in those cases entry is forced by ruthlessly demolishing protective barriers, or by cunningly bypassing alarm systems. Homes and offices of doctors are frequent targets, possibly because they are expected to yield narcotics as well as usual valuables. In some homes, fully protected with excellent door locks, window gratings, and extensive alarm systems, thieves break in by cutting through the roof, or penetrating a wall from an adjacent garage. On occasion, residents are brutally pistol-whipped and tortured to open a safe or reveal the hiding place of their valuables.

Other Home Hazards. Concern with the growing crime threat should not divert attention from the ever-present menace of fire and other perils which exist within the home itself. Also, the present high cost of maintenance and repair services often leads to neglect or delay of essential upkeep, inevitably resulting in rapid deterioration of valuable property and the development of dangerous conditions.

One example of such combined damage and danger is a clogged roof gutter, which causes an ice dam in winter. Since the thawed water can't run off, it seeps into the eaves and dampens the walls and ceiling. Plaster, when wet, loses its bond to the lathing, and the ceiling ultimately may come down without warning in massive, heavy chunks, periling any person in the room. Roof gutters need regular seasonal cleaning. After correction of any roof leak, the firmness of nearby ceiling areas should be tested and any loose sections knocked down. In doubtful situations, the entire ceiling usually must be done over.

Fire Warning System. The most feared home hazard is that of fire, and with justification. Fires and explosions destroy or damage more than 600,000 one-and-two family homes and other dwelling units, and take about 6,000 lives annually in the U.S. A careful program of prevention, based on familiarity with the causes of fire, explosions, and burns, together with a dependable fire warning system, will ease your mind and protect your family and home. Details are given later regarding the selection and installation of fire alarm systems, together with data on the use of fire sprinklers in the home.

Other causes of serious injury in the home are falls due to shaky stair railings and encumbrances, structural defects, electrical shock, cuts, accidental poisoning, falling objects, and those resulting from careless or improper use of tools, appliances, and firearms. Knowledge of these dangers and alertness in spotting such hazards to persons and threats to the property itself, will help to eliminate them.

"Acts of God." When the disaster of a hurricane, earth tremor, forest fire, mudslide or tidal flood strikes, there's nothing much that can be done that had not already been done. Earthquakes can happen anywhere, but some sections are more susceptible than others. Cyclones and other major storms are even less predictable. The homeowner can profit from expert opinion, however, planning ahead to reduce such destruction as was caused by the mudslides that occurred in 1969 on the Pacific Palisades, the 1970 forest fires, or some of the Gulf Coast hurricanes.

Nearly all homes now are protected from lightning bolts by their grounded electrical system. In some situations, however, subsequent construction work may affect the original grounding effectiveness of the system, and a periodic checkup is advised. The basic principles of this grounding system, and of the lightning rod, are explained in Chapter 17.

Environmental Invasions. Another condition only recently receiving the attention it deserves is the invasion of residential areas by industries, institutional facilities, and even municipal plants that pollute the area with obnoxious odors, cause excessive noise and traffic even at night, accumulate huge piles of trash, or blanket the neighborhood with smoke, dust, or radioactivity. Homeowners and residents need not yield to this imposition in subdued resignation. Suggestions are presented on ways to ward off a threatening or even existing condition, ways that have in many cases been successful in protecting the neighborhood. Similarly, property damage to individual homes, such as cracked foundation walls done by blasting for nearby construction, may be recoverable by proper measures.

How Much Protection? Attitude is a vital factor in shaping your defenses. You should recognize that there is no such thing as absolute and total safety—the only place that comes near that concept is a prison cell with constant guards. So whatever provisions you make for your personal and home protection, there must necessarily be a compromise between total security and the requirements of normal daily living. Any program that fails to meet this standard indicates a surrender to fear.

You wouldn't want to live behind barred windows just to avoid a possible intrusion. Nor would you choose a watchdog to protect you that is so wild and ferocious that it might maim or kill an innocent neighbor—or even turn against you at some provocation.

The important thing is to strike a balance, accurately defining the menace, and proportioning the measures necessary to defeat it. The weight of this presentation is that most burglaries and home accidents are made possible by carelessness and neglect; and that careful adherence to the *Habit of Safety* can give you the best chance to avoid being victimized by the wave of crime.

Large companies and industrial plants employ security consultants to plan controls against thefts, intrusions, and other crimes. Such professional services are not usually available to individuals. This book is intended to fill the gap so that the sanctity, the privacy, and the safety of the home can be maintained. Readers often can obtain advice on specific problems from some of the companies producing or selling burglary protection materials, but it is necessary to caution you against self-serving recommendations. A list of sources for useful booklets is included at the back of this book.

2

No Admittance!

The elements of a home defense program may be classified as follows:

1. Avoid attracting the attention of housebreakers.
2. Provide adequate barriers to entry.
3. Keep an alarm system or a watchdog on the alert.
4. Set up a security room for family retreat.

These four elements for burglary protection all depend on one underlying factor: the Habit of Safety, which is the human factor that forms the cornerstone of your efforts, and on which success depends.

Prowlers Keep Out! The first objective of your program is to avoid attracting the attention of a potential burglar to you or to your home. Whatever the extent of your affluence, there's no glory in flashing your diamonds while shopping for the family groceries, or pulling out a big wad of bills to pay a car repair bill. Instead, pay by credit card or charge account whenever possible. At your home, let delivery-men wait at the door, or allow them only into the kitchen if necessary, so that as few people as possible get a view of the house layout, particularly the location of the burglar-alarm control box. You might be confident that your local tradespeople do not have dishonest employees, but youngsters have a tendency to describe an impressive home setting to friends or schoolmates whom you do not know.

Even for obviously prosperous homes, it's a good policy to keep the family Cadillac or Continental, and that sporty runabout, out of sight and in the garage. Luxurious lawn settings and swimming pools might well be screened by a high board fence or shrubs; they won't keep a burglar out, but they're less likely to invite him in.

There's no need to disguise or hide your protective measures, however. Evidences of well-planned security such as an alarm box high on the wall, a red alarm indicator light, steel gratings over basement window wells, and well-polished solid-looking door locks may effectively discourage a prospective intruder.

Thieves go primarily for money. They'll look for jewelry and other articles, but usually only if not enough cash has been found. They know that their risk is greatly multiplied if they are caught with "the goods," and at best they'll have to dispose of the stolen articles for very little money, regardless of their value.

A good rule for the homemaker is never to accumulate large sums of cash at home. That's what banks are for. It's safer and more efficient to keep on hand only such amounts as are needed for everyday purposes. Elimination or reduction of cash hoards, if that were generally adopted, could well bring a downturn in the number of burglaries.

But don't go to the other extreme and strip your house of all cash. There's an old folk lore, tried and proven for many decades and still valid, that it's wise to leave something for the crook. Empty-handed burglars can become so mean and frustrated that they go on a rampage of malicious destruction, slashing upholstered furniture, ripping clothing, smashing chairs, even deliberately blocking the sink drains while the water is turned on full force. The simple solution often is effective: keep $10 or $20 in an obvious place where it will

The sensible precautions pointed up in the sketch can become part of the family's routine. Check them off for your home by the numbers on the sketch. 1. Keep garage door always closed, with a lock as secure as the one on the front door. 2. Make sure front and back doors have night deadbolts in addition to spring latches. 3. Outside lights are a strong deterrent to burglars; there should be one covering every point of entry. 4. Don't keep large amounts of cash or valuable jewelry in the house. 5. Basement entry should preferably be covered with steel doors; padlock hasp should have all screws covered when locked. 6. Shades should be left at usual level when family is away, not tightly drawn. 7. Basement windows are covered with heavy wire mesh. 8. Plantings are trimmed so they won't screen a burglar's entry. 9. Don't hide the door key under the doormat, or on the door frame or in the mailbox. 10. Stop deliveries when the family is away. 11. A ladder is best locked with a chain or stored in a locked place.

be found. Loss of a small sum is insignificant if it prevents such vicious and costly vandalism.

No Easy Pickings. Home safety is not to be had just by wishing or hoping. It takes time and effort for planning the protective program, complete cooperation from all members of the family in carrying out the necessary routine, and also the expenditure of a modest amount for whatever hardware is required to do the job. Pennypinching thrift that postpones action often has a disastrous sequel.

15

Making a Security Survey. A first step in the security program is to make a careful and complete inspection of your home premises to spot vulnerable points of entry, judging them always from the point of view of the possible prowler. Carry along a small note book to jot down your findings for correction. Are there clumps of bushes around the side or rear door that would provide convenient cover while a thief is forcing the door or breaking a window? Perhaps trimming the bushes lower or aiming a flood light at that location will correct the situation. Does a large tree limb extend close to an upper porch deck or open window? Pruning the tree will eliminate the problem. Is the side door of the garage left unlocked, enabling an intruder to hide there while watching the movements of the household? Are your telephone wires exposed so they can be clipped easily, cutting you off from help? The wires can be protected with conduit, or shifted to rooftop level. Does your dog bark habitually at night, canceling his value as a watchdog? Does every exterior door with glass panels have a double-cylinder lock which requires a key to open it from the inside, and is that key kept at an inaccessible distance from the door? Are all ladders on the premises secured with padlocks or otherwise out of reach? If you have sliding glass patio doors, do they have a Charley-bar or extra throw bolt, as illustrated, so the door can't be lifted off the track? Are basement windows and similar means of entry protected with metal guards or areaway gratings? Are your special valuables kept in a vault or well protected closets, and do you keep a list of their identifying serial numbers?

TO SUM UP: a home is relatively safe if all doors have solidly mounted dead-bolt locks, the windows are all closed and latched at night, the rear yard adequately lighted, basement windows and glass panel doors protected with metal guards, all keys strictly controlled and accounted for, callers always identified through the viewer before you open the door, and there is an effective burglar alarm system that is kept on alert, or a dependable watchdog. This routine is not particularly difficult, and soon becomes established as a habit of safety. But a single door or window left unlatched, or the one time

Sliding patio doors must have an adequate locking device. A horizontal "Charlie" bar across the closed door prevents easing the door out of its tracks.

the alarm is not set, cancels all your efforts and is an invitation to burglary.

Apartment occupants have especially difficult security problems and there is no intention to slight them here. The words "home" and "homeowner" as used throughout this book refer as well to apartments and apartment tenants. A separate chapter is devoted to apartment security because of the special problems faced by the millions of apartment dwellers.

Selective Purchases. Spurred by incidents of housebreaking in the neighborhood, homeowners rush out to the hardware store for additional locks and door chains, buying whatever types are available. Often these purchases are influenced by extravagant claims for "pickproof" locks.

Responsible dealers usually will remind you that even a very good lock is no better than the door on which it is installed. Burglars rarely bother to pick a lock unless it's an easy type with a well-worn keyway. Most often, a door is forced by the leverage of a jimmy or heavy screwdriver inserted between the door and jamb. A lock, any lock, won't defeat a really determined burglar, but a good dead-bolt lock in a solid door jamb can make breaking in so slow and noisy that the thief may give it up as a bad job.

Door chains are fine, ever so good-looking in polished brass and trim lines. But these are sold with tiny screws that won't offer much resistance to a determined thrust. Not only are longer screws advisable, the chain should be mounted in such a way as to provide minimum leeway or leverage to an attacker. Enough opening to identify a caller, or receive a letter, is sufficient.

3

Who Is the Burglar?

Knowing the true nature of the enemy is helpful for setting up defenses to balk him. What kind of person is able to prowl around homes and apartments without creating immediate suspicion? What type will take up this illegal, immoral, dangerous, and poorly paid activity? Are they skilled Jimmy Valentines with sensitive fingers that can open combination safe locks by touch? Are they bumbling and ignorant brutes who simply demolish doors by massive strength? Are they dope-crazed addicts indifferent to any consequences of their acts? Or are they just callous youths preying on the unwary for pocket money?

A Composite Picture. A fairly conclusive portrait of today's housebreaker might be put together from arrest statistics and the data from various studies and surveys. The composite picture that emerges is of teenage boys and youths, 15 to 21 years of age, living in the neighborhood or nearby vicinity, school dropouts without jobs or skills, very likely taking narcotics, of lower socio-economic background, and with a police record of juvenile arrest but not jailed.

Thrill Burglars. With more than a million burglaries, many of course are committed by persons who do not fit these characteristics. A large number of housebreakings are done by youths from affluent homes — even the son of a Governor (Maddox of Georgia) was arrested twice for burglaries. Apartment thefts often are the work of tenants in the same building who are short of cash and have time on their hands.

A small percentage of the crimes may be committed by older persons, some of them hardened criminals with long prison records, who know the consequences of getting caught. So-called professional burglars rarely go for haphazard marauding, house to house, ready to tackle anything at any opportunity. Rather, they are more selective, will scout a likely target carefully for the possibility of a large haul in cash and jewelry or art objects, and with minimum risk. Some of these ex-convicts have extensive acquired skill in picking locks, bypassing burglar alarms, crashing through formidable barriers, breaking window glass or removing glass panes intact from their frames without making a sound, and even trucking away heavy safes from homes. But they know that the odds are against them, that arrest means a longer jail sentence than the last time, so they limit their thefts, unless they are desperate for a quick stake. While 77 per cent of burglary arrests are repeaters within a 5-year period, it is the younger group that has the highest record of recidivism. But not all burglars are hit-and-run teenagers, eager for small pickings. Some are men with occupational skills, unemployed, or unable or unwilling to hold down jobs, and preferring to exist as thieves, traveling from town to town.

Suspicion Avoided. Some have extreme daring and aplomb, growing out of long experience. They do not act surreptitiously so they don't excite suspicion. Some even operate with a small, neat truck, usually one rented right in the town or perhaps painted with signs to indicate that it is a delivery or service van, and even wear a sort of standard deliveryman uniform. They roll right up to the door (after making certain no one's at home), make a quick entry unobserved, then calmly move out whatever they think valuable — television sets, audio systems, even rolled up Oriental rugs and good funiture pieces. If caught in the act, they have some ready-made explanation that often gets them free.

In one instance, in a neighborhood where hardly a home had escaped being burglarized, families of relatives lived on the same street. One woman was outside wheeling her infant's carriage and talking with friends when she noticed a truck in front of her brother's home down the street and saw two men carrying out articles of furniture. But she recalled her sister-in-law mentioning that she planned to have some furniture reupholstered. Early that evening, however, her brother and sister-in-law excitedly phoned the news — their home had been emptied of half their belongings, right in the daytime, with an ordinary-looking truck, and with a relative looking.

Analysis of Arrests. Burglary arrests in 1970 totaled 358,100 persons, and reportedly solved about one in five of the burglary cases. The figure does not include many arrests of juveniles, since those records are not reported by certain community agencies. Of the reported arrests, about half were in large cities. Suburban areas reported 67,700 such arrests, while the rural figure was only 21,500.

Of all these arrests, about 65,000 were under age 15, and 83,000 were ages 15 to 17. Including non-reported juvenile data, more than half of the burglary arrests were under age 18. In contrast, arrests of all suspects over age 25 totaled about 17 per cent.

Brazen Gutter Swipes. Thieves don't even have to *enter* a house to accomplish their purpose. They'll take anything of value on the outside, whether it's nailed down or not. One prime example that has been reported from all over the country results from the current high prices for scrap copper. The vandals just come along and rip off whole copper gutters and leaders, fold and stamp them into a flat bundle, and off they go. The astounding thing is that as scrap the copper may bring perhaps $25 in a junk yard, while the replacement costs would run as high as $500. These gutter swipes often operate in broad daylight, don't even need a ladder to scurry up and loosen the gutters with a crowbar, and are rarely caught in the act since they are usually taken for workmen. At least one instance has been reported where this outrageous thievery occurred while the family was right at home.

Valuable shrubs also are the frequent target, and it's not because the thieves are nature lovers. Many fine shrubs bring $25 to $50 each, and these youngsters are known to actually take orders, particularly from new homeowners who require considerable plantings. The vandals then just wander along to find likely items, and quick as a wink they've denuded some lovely front lawn or garden.

Protection against such looting is difficult to provide. Fences may provide some safeguards, but a good watchdog is probably the only reliable defense.

Loot Brings Little Cash. The amazing thing is that the burglar usually winds up with so little for his efforts, and for all the risk he takes. The average reported loss per burglary is about $330. But that figure usually is based on original cost of the stolen articles. What does the thief get? In the "no questions asked" channels in which the goods are unloaded, typewriters might bring as little as $3, mink coats or a chest full of sterling silver will go for perhaps $20, watches and bracelets command even less, and diamond rings will bring no more than a tiny fraction of their value—if the burglar knows the real value, which usually he doesn't. In one case reported by police, a "fence" paid only a fraction of the melted down value of fine antique silverware. Television sets, adding machines and cameras are peddled in the back streets for $10. Spare tires, stolen on order from the trunks of late model cars, net only $5 or $10. Legitimate pawnbrokers won't touch any of this stuff.

So the burglar, for all his pains, gets very little of his actual haul unless he stumbles on a cache of cash money in the house. If he does, the chances are that he won't bother carrying off any large articles because of the extra hazards involved.

On an average, the payoff for a highly dangerous housebreaking job may be as little as $10 or $20. Which prompts the question: Why do they do it?

Many are Addicts. In New York City approximately 10 per cent of all persons arrested admit to narcotics addiction — the actual number would be considerably higher, but only voluntary admissions of addiction are entered on the record. Almost a third of the admitted addicts are in the 16-to-20 age group.

Experts have stated that the addicts, "junkies," require up to $60 a day for the drugs. Many addicts are unfit or unwilling to hold regular jobs, and would not earn enough even then, so they resort to any means of securing money — even just enough for one more day's supply of the drug. Hence, burglary becomes the way out. In congested city core tenements, junkies have been known to break into a neighbor's apartment and cart away the entire contents — clothes, bedding and all — to get only enough money for the next heroin "fix."

Girls and women are increasingly involved in burglaries. Although the number is relatively small, a possible explanation is that the females operate in concert with men, acting usually as lookouts. However, some sensational cases of women burglars have come to light, such as that of a housewife, 26 years old, who police say admitted burglarizing more than 200 homes in the Detroit area during a 10-month crime spree. She was designated the "pillowcase burglar" after police found 25 pillowcases containing $200,000 in loot secreted in her attic.

A different type of burglar was revealed when police broke up a gang of five "cat burglars" charged with stealing $20,000 in more than 100 burglaries over a period of a year — an average of two a week! The members of the gang were 14 and 15-year-old boys, all from respectable middle class and working class families living in the neighborhood they preyed on. In the same area, 17 youths ranging from 15 to 23 years old, from prosperous families, were previously arrested and charged with a series of 500 home burglaries.

Almost any door can be forced open if a pinch bar can be wedged between the door and the jamb. Pressure is exerted to pry the lock bolt out of its retainer plate. A door with two sturdy locks near top and bottom, is much more difficult to force.

Crowbar Main Tool. The main tool in the burglar's hand, often his only one, is a pinch bar — the "jimmy" — known also as a crowbar, pry bar, rip bar, or wrecking bar. Often a heavy duty screwdriver serves the same purpose. In 12-, 15-, or 18-inch lengths, the slender bar is carried unobserved inside a pants leg, suspended by a cord from the belt. Forged of high-carbon steel, with flattened ends bent at a slight angle, the bar can exert tremendous leverage to force a door bolt out of its latchplate, bend steel window casements, or lift heavy iron gratings. One chisel-shaped end of the bar often is filed down to form a thin blade that can be forced into any narrow space. The other end, which is forked to use as a nail puller, provides a purchase for ripping out locks or chain hasps, and similar efforts. The pinch bars sell for as little as $1.50.

How Burglars Operate. Burglaries follow standard patterns that are too similar to be coincidental. In areas of private homes, one or two couples in a car parked unobtrusively under dark trees closely watch the comings and goings at nearby homes. When the occupants of a house douse the lights and get into their car, obviously leaving for the evening, that house is marked for attack. The couples cruise around the block several times to make sure all is clear, then the men slip out and dive into the driveway or behind bushes. The driver parks nearby, awaiting a whistle or flashlight signal to pick up the burglars and their loot, and off they go.

Sometimes youngsters hidden in backyard bushes, or back of the garage, give the signal to their leader, who checks over the premises and selects the point of attack. As entry is forced, the accomplices place garbage cans or other obstructions in the driveway to give notice of any arrival. As soon as they get inside, their first steps are to establish secondary escape routes by opening wide one or more windows, and to block all entrances by putting on the door chains, throwing the barrel bolts, or if these are lacking, piling chairs or other furniture against the doors. That is why police arriving at the scene do not rush into the house but rather circle around to head off escape at the rear.

If an alarm is set off, the burglars make a hurried effort to find the alarm control box and clip any exposed wires or pull the electric fuses. Sounding of the alarm does not always cause a panicked escape, since prowlers have learned to play it cool knowing that neighbors may not pay attention. But the clanging alarm does cause a more nervous, hurried search.

If things remain quiet, the hunt begins for the hidden cache of money or jewelry they hope to uncover, shades are carefully lowered before any lights are put on. Where to look? All the favorite hiding places are known: cookie jars, inside hat boxes high on closet shelves, under piles of shirts or negligee in drawers, on panels under or behind the drawers, or taped to the back of pictures. Everything is hurriedly ransacked, turned topsy turvy, while a lookout at the front window watches for arrivals or a car horn signals danger. The loot is placed in paper shopping bags or small suitcases. In most cases they will take only cash or expensive jewelry, ignoring watches and other identifiable items that would bring them little money.

In apartments, a burglar entering from a window would immediately put on the inside door chain to prevent being surprised by arrival of the tenant. If entrance is made by forcing the door, the fire escape window is immediately opened for escape, and the door itself barred.

These details of burglary patterns point to some defensive measures.

• A car with occupants parked for any length of time on the street calls for investigation, particularly if the occupants seem trying to avoid attention. It is a difficult matter to call police every time one suspects a car. Still it is important to maintain an alert for a possible danger; the police are always ready to cooperate and know their business.

• When leaving the house, do not put out all inside lights, which would indicate that no one is at home. Both interior and outside lights are useful.

• When returning home, if there is a strange look about the house — cans in the driveway or shades drawn when they should not be, do not stop but continue to drive on and telephone police from another location, describing the situation. They will know what to do.

BURGLARY BY THE NUMBERS

In the state of New Jersey, "Breaking and Entering," which covers all forms of "burglary," is by far the most "voluminous" of the Crime Index offenses, and the statistical details for a recent year are very close to the experience of other states.

Total breakings and enterings:	71,445
Residence:	38,788
Non-Residence:	32,657
Increase over previous year:	18%
Proportion of burglary to all crimes listed:	41%
Nighttime burglaries of residences:	14,584
Daytime burglaries of residences:	16,384
Robberies in residences:	831
Total dollar loss:	$16,500,000
Average loss per burglary:	$423
Entry by force:	80%
Entry without force:	11%
Unsuccessful attempts at entry:	9%
Increase in daytime home burglaries, in one year:	30%
Police solution of burglary cases:	8%

Age of arrested suspects: more than 50% under 18.

4

Watching Your Step:
The Habit of Safety

Whatever security devices and barriers you have installed, everything still will depend on how fully the Habit of Safety becomes part of the pattern of living for every member of your family. Even the finest lock won't do much good if the spare key is hanging on a nail in plain sight of a prowler in the garage, or a duplicate has been made by some part-time helper in the car wash where you left all the house keys in the ignition lock. What kind of security does a kitchen screen door provide if all the family is watching television in the den?

Home security takes a price, not so much in money, but rather of constant alertness and adherence to specific safety rules, even if they require some daily inconveniences and changes in long-established routines.

Who's There? Everyone knows how important it is not to open the door until the caller is positively identified, through the viewer or intercom, as *someone who is known and trusted.* Yet time after time, serious incidents occur because of carelessness about this fundamental rule. That certainly makes it easy for criminals: why take all the risk and bother about breaking in when people accommodatingly open the door.

A set of rules should be established at your home, to be followed by every member of the family. Use the viewer to scan the stoop or hallway, and refuse to open unless the caller is known. If the door must be opened, first put on the chain so that your observation can be confirmed. These steps are necessary and justified, so don't feel that you will be considered fussy or overcautious.

Make sure the outside light is bright enough so that you can observe the caller. This may be a problem in some apartment corridors where lighting is provided at minimum level. Local ordinances in some cities now require adequate illumination in apartment house hallways and lobbies, in addition to spotlights at service entrances. Even so, vandals have made a practice before invading apartments of breaking the lamps to assist their getaway and hinder identification. Your suspicion should be aroused whenever the lights are found to be sabotaged. Notify the "super" by telephone, and try to remain inside the apartment for the evening, or until the condition is corrected.

If your door viewer doesn't cover enough of the outside area, so that someone standing at the side can't be seen, get the improved swivel-type viewer that permits sighting side to side, and even downward. One such viewer is sold by Ajax Manufacturing Co.

You even have to be watchful when entering your home that there is no one loitering or hiding nearby who will simply slip inside just as you unlock the door, holding a sharp knife against your body to keep you silent. And silent you should be under the circumstances—just hand over your cash and hope the intruder leaves.

Phony "Home Buyers." With the more frequent daytime burglaries, a favorite recent trick for gaining admittance is worked by a man-woman team posing as possible home buyers. A For Sale sign is like an engraved invitation. The pair can roam the house, spotting the possible locations of money or jewelry, making sure no one else is at home, and sometimes

Eye-level lens gives wide-angle view of entrance area. The rotating viewer shown covers greater area than stationary type.

even taking the precaution of clipping the telephone wires or burglar alarm system. Then, they make quick work of grabbing whatever they can and make a getaway in a car parked around the corner. Listing a home for sale through a legitimate local realtor can help avoid such dangerous invasions.

Another gimmick is to lift a woman's pocketbook in a department store, then telephone to tell the victim that her bag had been found intact and that she should return to the store to claim it. As soon as she leaves the house, the thieves are waiting to enter with the keys found in the bag.

The decline of Western Union may be a blessing in at least one way, the virtual disappearance of the messenger boys and with them that old "telegram delivery" trick of gaining admittance. But in its place has come the "package for you" or even that real oldie, "flower delivery" that few seem able to resist. If you insist on opening up in these instances, keep the chain on while you obtain and sign (and inspect) any necessary receipt, then instruct the deliveryman to leave the package outside the door as you "can't take it in just now."

Installation of Peek-O-Viewer requires drilling a hole through the door. Locate a position that will be clear of any obstruction, mark center with awl.

A hole saw will drill the required 2 ⅜-inch hole. If necessary use a 2 ¼-inch hole cutter and enlarge the hole with a rasp. Start the pilot drill in the marked hole, hold the drill straight in and drill to full depth of the hole-saw cup, which will be about half the thickness of the door.

The pilot bit penetrates the entire door thickness, thus providing the guide for starting the drilling with the hole saw from the inside surface.

Cutting completed, core is removed intact with the hole saw.

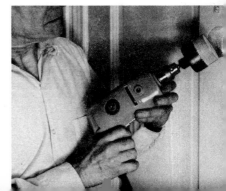

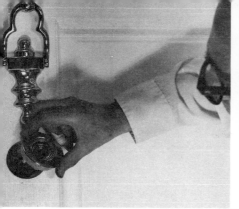

Rotating viewer section is inserted through exterior door surface.
Plate on interior side locks the assembly securely together, and the
installation is completed.

More Daytime Tricks. Perhaps these daytime hoodlums
are getting more shrewd, as demonstrated by this incident: A
young man rings the door bell and announces that he had
"just damaged your car outside" and wants to leave his
name so you can send him the repair bill. This happened at
a home of a very quiet and careful family. When that reasonable statement opened the door, a couple of thugs rushed in,
accompanied their demands for "your diamonds" by seriously
injuring one of the family. In this case, the door viewer was
used, but the second man had remained pressed against the
side wall, out of the viewer's visibility range. Had the door
been kept on the chain, even though opened, entry would
have been prevented. As it was, some $20,000 in diamond
rings and cash was taken.

Your Defense Plans are Secret. While there's no purpose in covering up the fact that your home is well protected, at the same time you must keep the inside arrangements as "classified secrets." The fewer people the better,
who know the location of your alarm cutoff switch, what types
of detectors are used, whether you have an open or closed
circuit alarm (very important information for skilled burglars)
and other details. Particularly with delivery boys and service-
men — the right thing is not to leave them unattended anywhere in the house. For example, snipping the open-circuit

31

Chain with lock can be attached from outside when leaving the apartment.

wire takes just a second and negates your entire alarm system. A duplicate key or bending of a detector contact of a door alarm means open sesame to an intruder.

Hold on to Your Keys. There is much misunderstanding of the ways in which keys can be duplicated. It is not necessary for the person to have the key long enough to get the duplicate made with a keymaking machine. All he needs is an impression of the key in candle wax, plaster, or other suitable medium—then the copy can be produced from the correct key blank by filing the notches indicated by the depression. Duplicates can be made also from the numbers stamped on the key.

Changing Lock Cylinder. Any suspicion that a key is lost or missing, or may have gotten into wrong hands, should prompt immediate change of the lock cylinder. This is a simple matter, particularly if you keep extra cylinders with their keys on hand. Perhaps the cylinder you take out now can be put back again at some future date when another change is indicated, so save it with its set of keys. A more thorough measure would be to have the cylinder code changed each time, with new keys made, as described in the lock section.

One extra detail: keep a record of the serial numbers and descriptions of your valuables. It's a good idea to place a hidden identifying mark on jewelry, watches, bicycles, cameras, typewriters, etc. Photographs also may be useful. If there's a loss, such information will be most helpful to the police, and may be your only hope of recovery or of collecting on your insurance.

HOME SAFETY CHECKLIST

1. Never open the door until the caller has been positively identified by sight, through the viewer or chain guard.
2. Put the door on deadbolt lock every time you leave, even for short periods. Older locks require use of the key to move the bolt, but some new types have automatic dead-bolt setting.
3. Keep a tight rein on your keys. Detach house keys from the case when leaving your car for service. If a spare key is hidden outside for emergencies, select a place that won't be discovered, and keep it secret.
4. Don't depend on the screen door and screen windows, even in the daytime. When leaving the room, close and lock the prime door and windows.
5. Have someone at home to receive children returning from school, or arrange with a neighbor to let them in, rather than providing them with keys which can go astray.
6. Keep garage doors closed and locked at night; also lock car doors even in the driveway.
7. Light up the side and back yards at night.
8. Always put burglar alarm on alert at night, and whenever you leave the house.
9. Prevent access to your tools; place padlocks on ladder brackets.
10. Nighttime lockup includes all accessible windows.
11. Put an identifying mark or number on all portable valuables—jewelry, TV, cameras, bicycles, appliances, even furniture.

CHECK LIST FOR ENTRY BARRIERS

1. Solidly installed door locks with deadbolts of sufficient throw to deeply engage their latchplates. Doors must be in good condition, fit snug in their frames, with no space between the door edge and jamb to admit a screwdriver blade. Always double-lock the doors when leaving, as the spring latch alone is an easy mark.
2. Doors having glass panels, or with glass lights alongside the frame, are particularly vulnerable. Resurfacing with plywood panels to cover the glass may be advisable. Otherwise, use double-cylinder locks, requiring a key for opening from the inside. The key should be kept out of the lock, in a handy place that is not accessible by reaching through the opening.

Lightweight kitchen and basement doors of many homes are poorly protected with a simple mortise lock that is operated with a "skeleton" passkey. Supplementary throwbolt or doorchain is little help on a door with glass panes.

An unsafe door lock can be replaced quite easily with a rugged and efficient lock that provides double security—a slip latch that cannot be retracted because of a recessed retainer pin, and a solid deadbolt. Heavy security plate protects lock cylinder from tampering.

Door chain permits partial opening of the door to interview callers, receive deliveries, while retaining adequate protection against unwanted entry.

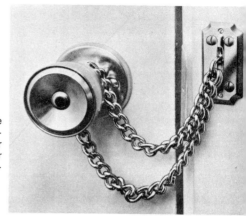

Stronger and more convenient to use than other types of chains, this one simply slips over the door knob. Called Door Guard, it is mounted to the jamb with four extra heavy screws. Made by Ajax Hardware Company.

3. Doors opening outward, such as found in some apartment houses, are hinged so the pins are on the outside, where they can be removed easily. The door then can be opened on the hinge side (there is sufficient leverage to overcome the short deadbolt in the latch.) Special hinges are available with pins locked in place by setscrews which are accessible only when the door is open.

4. Door chains are dependable if the chain plates are installed with sufficiently long screws that can't be yanked out by swinging the door. The lock-and-key type of chain, which can be set when leaving the house, is useful, as are barrel bolts. There also are special heavy duty steel bars and braces for doors that are difficult to defend.

5. Patio sliding doors generally have inadequate locks and may be nudged off their tracks if there is sufficient movement to clear the jamb pocket. A hole drilled through

A surface-mounted throw bolt provides added security, particularly against possible possession of duplicate keys by unknown persons. This Stanley night bolt has extra strong jamb plate.

Spring latch is a useful adjunct to the standard patio door catch, preventing the door from being lifted out of its track.

both door and channel for a heavy nail or pin can serve as a double lock. Also useful are Charley bars, or spring catches.

6. Windows at porch decks, on ground level and in basements, are difficult to protect. One simple measure is to drill small holes through the sash for nails that prevent raising the window. Various hole positions can be planned to allow opening the window part way for ventilation.

7. Built-in window locks prevent raising the sash even when the glass is broken. Keep the key in a handy place for emergency use. A more effective protection is obtained by glazing the sash with double-thickness or mesh-reinforced glass, or with thick transparent acrylic plastic which is difficult to shatter.

8. The ultimate protection for windows is some form of steel guard or gate. Basement areaways may be closed off with heavy gratings, or bars fixed into the window frames. Apartment windows accessible from the fire escape or terrace may not be blocked off or encumbered, under the fire regulations, but some forms of metal guards or folding

Vulnerable windows can be protected with steel grilles, fastened securely to the window frames. However, exit from windows facing fire escapes must not be blocked. In those locations, window guards may sometimes be used with a secret but easily opened inside latch. Make sure installation conforms to local fire regulations.

gates may be used providing they are fitted with an easily released catch. A local firm specializing in these window guards may be able to offer acceptable types, but don't compromise with fire safety to prevent burglary.

9. *Comeback for Window Shutters?* Long relegated to strictly decorative uses, the old-style window shutters may be due for a practical comeback as their value for window protection gets renewed recognition. Even in ground-floor windows, and those in apartments facing fire escapes, shutters can be a powerful barrier to both prowlers and voyeurs, while allowing adequate light and ventilation by means of movable louvers. On the European continent, particularly in Italy, France, and Switzerland, the shutters are in constant use for both privacy and protection.

Ready-made, low cost shutters usually can be cut down to fit the opening precisely. The shutters are mounted outside in pairs on hinges or swivel-type brackets, or inside the window frame. While it would appear that the outside shutters could be removed by lifting them off the brackets, that is so only when the shutters are open and folded back against the wall. When closed, however, they overlap to fit into the window recess, and thus cannot be removed short of splintering them apart. The shutters are closed on the inside with a simple latch, and they can be fitted with a burglar alarm detector for additional safety.

Interior shutters, those fitted on the inside window frame, are quite attractive and can be harmonized with the room decorations.

THE ELECTRONIC WATCHMAN

A burglar alarm system is an essential element in any home protection program. 1. The mere existence of the alarm system may deter attempts at entry. 2. If entry is accomplished, sounding of the alarm may scare the burglar away. 3. The alarm signal alerts you to the danger of an intruder in the house. 4. The alarm may bring help from neighbors or the police, and 5. Setting the alarm at night lets you know whether you've missed locking a door or window.

Obviously, the unit will do no good if it is inoperative,

either through some defect in the system, or failure to switch it "on," and may actually be harmful if you rely on it and have let down your alertness.

Thus, if a member of your household comes home later than the others and fails to reset the alarm on entering, the system is non-functional. In some instances, neglecting to set the alarm when going out "just for a little while" has given the burglars all the chance they were waiting for.

Testing is Essential. Regular tests of the alarm system are important to make sure all is in working order. The system may be in place for a year or two, even five or more years, before it is put to the real test of signaling a forced entry. Meanwhile, the tendency is to occasionally neglect throwing the "on" switch, possibly in resentment at adjusting one's actions day after day to the sensitivity of an electronic device.

A good system will have a means for easy testing of all its circuits without sounding a false alarm. Closed circuit systems nearly all have a constant "all's well" indicator lamp—even so, the bell or siren should receive a quick sound test periodically. Open-circuit systems can be checked out with a test button.

Power source is an important element in the system. The more sophisticated systems function on transformer current from the house power, with a 6-volt standby power pack, which automatically switches on if the house power fails. Other systems rely entirely on either transformer or battery power, and are thus subject to failure, or can be quickly silenced by a knowledgeable burglar.

Small, self-contained one-station alarms for apartment doors and similar locations use either a pair of 1½-volt dry cell batteries, which have a "shelf life" of 6 months to one year, or the small pencil-type batteries which have a more limited service span. While there have been improvements in design of these batteries, there always is the possibility that the alarm may malfunction when needed because of battery failure or corrosion of the terminals. Periodic replacement of these batteries is, therefore, essential. Some of the new security locks have similar self-contained battery-powered alarm units.

False Alarms Harmful. In any house with a newly in-stalled system, the door may be opened in a forgetful moment while the switch is "on," and result in a false alarm. Repeated happenings of this kind can nullify the value of the system, since neighbors will no longer respond, and even the harassed police will ignore it. Some systems, particularly the closed-circuit type, are so sensitive that they cause frequent false alarms, usually in the middle of the night, to the annoyance of everyone. This may be caused by vibration of the doors and windows by high winds, or simply the "dropping out" of a relay. Such events soon lead to disuse of the system, and de-feat its purpose.

Photoelectric and vibration detectors are subject to sounding off without reason. The open-circuit system, how-ever, is much more simple in its circuit design and therefore less subject to false alarms.

The Family Security Retreat. What do you do when the alarm signal informs you that there's a burglar in the house? Do you rush forth into the darkness to do battle? Never! You gather the family into the room that has been set up as the retreat for such situations, lock the door securely, then try to obtain help. If you can put on the house lights before you enter the room, so much the better. If there's a telephone in the room, call the police, or shout to neighbors. Otherwise, just sit tight until you are sure all's clear. But you're ready with an escape method, should the intruder at-tempt to break through the inside door. If that occurs, you would still have precious moments to get your family out the window and down a rope ladder to safety, with the burglar still in the house. Details of setting up the security room are given in Chapter 8.

5

Good Locks:
The Key to Security

Door locks are on the front line of your home security, so like most homeowners and apartment tenants you've probably worried about whether the locks on your doors are strong enough, safe enough. Perhaps you've recently changed or added a lock or two to bolster your feeling of security, but still are uneasy about the protection they provide. You've been impressed, no doubt, by what you've heard of recent technological advances in lock design and wonder whether these new locks will do a better job for you.

It is true that major improvements have been made in lock designs recently that incorporate many advantages. New concepts of cylinder mechanisms and unique key characteristics, together with more rugged construction, make these locks more resistant to picking and tampering. Stricter control by manufacturers over the key blanks make duplication far less likely. A valuable development also has been made in the formerly vulnerable spring latch by the addition of an automatic interlock that effectively forestalls the "celluloid artist" who can easily manipulate the ordinary bevel latch.

There are, in addition, a number of special purpose locks that meet the challenges of high crime areas or the need for extraordinary precaution. Among these is a bar lock, consisting of a pair of solid steel bars that slide into rugged hasps at each side of the door, a reminder of the heavy timber placed across the gates of a medieval fortress to withstand attacks with battering rams manned by hundreds. Another is a heavy rod braced against the door in conjunction with an ingenious lock, a modern-age version of the old chair wedged under a door knob.

Very likely, one of these better locks would contribute much to your safety and peace of mind. But the locks you now have may be of adequate quality, needing only a new cylinder to make them perfectly functional, or perhaps just a change of the pin tumblers for a new set of keys to give you a fresh start. Sometimes, just soaking the cylinder in a solvent bath like varnolene will restore it to proper function by washing out the grit and grease that may "freeze" the pin tumblers in the cylinder core. Another excellent step, perhaps the most effective one, is adding a second lock to each door—this doubling up has a definite value as it both reinforces the door and is one of the most positive defenses against picking. Especially important—if you don't have a good deadlatching lock, so that you know the door is always secure without further attention, that's something you need right away. The increasing incidence of daytime intrusions makes this a must.

Time for Evaluation. Before taking any further steps, it's best to evaluate the protection afforded by your present locks. It may be that some changes will be indicated, but possibly the lack of security may not be the fault of the locks you now have, or at least only partly so, and corrective action may involve factors other than the lock itself.

Have you ever thought about how a lock holds the door? You know, of course, that the lock "throws" a solid bolt that engages in a metal latch plate recessed into the door jamb. But in most common locks, the bolt moves hardly a half inch, usually even less. If there's a considerable gap between the door edge and the jamb, the bolt may be engaging the plate to a depth of only about an eighth of an inch—all that holds the door in place! It's hardly an adequate barrier, and no wonder that many a burglar has broken into a home just by giving the door a hard push. A properly fitted door, allowing just $1/16$-inch clearance at the jamb, combined with a quality lock that has a bolt thrust of three-quarters of an inch, represent a solid barrier that can hold fast against routine attack.

Another important detail is that the door jamb is usually only wood one-half inch thick, held with a few long finishing nails into a 2 x 4 wall stud that may be inches away. The latch plate is recessed perhaps an eighth of an inch deep, but there's a deeper recess beyond the plate for the lock bolt (if it ever reaches that far). It is at that very spot, at the latch recess, that the door jamb is weakest. If a pry bar can be inserted from outside between the door and jamb, only moderate force would split the jamb apart and release the latch bolt. But without the leverage obtained with an inserted pry bar, a properly installed lock can resist tremendous force directly against the door.

Stop Molding Removable. Another source of vulnerability may be seen on the outside of your door. Is your door jamb flush across its entire width so the door edge butts up against tacked on stop molding, or is there an offset in the jamb? Surface stop molding is a fragile barrier that can be easily removed to expose the door-jamb clearance. Then, if there's space enough to slip in a pry bar or screwdriver blade, the door yields quickly and almost silently because of the side leverage.

Then there is the wavering strike plate, barely held by screws that have long since loosened. The mortise, enlarged many times to adjust for a sagging door, no longer holds the plate in line, so that often the lock bolt doesn't even engage in the plate slots, and rather catches in the narrow strip of wood between the strike plate recess and the edge of the door jamb, often just a fraction of an inch of wood. Prying the door back against its hinges will free the lock bolt from this recess.

Solving Gap Problems. Here are possible solutions to these particular problems: For the gap between the jamb and door, you may be able to shift the jamb closer to the door, perhaps by inserting a strip of plywood of the required thickness, or by driving wedges between the 2 x 4 framing and the jamb, after removing the facing trim on the door frame. Possibly the threshold also will have to be lifted and cut back a bit for more clearance, before it is replaced. Be sure that the jamb is kept plumb, and test frequently by closing the door to be sure that you retain sufficient clearance.

If the jamb cannot be moved, for one reason or another, it may be possible to fasten a covering strip over the present jamb, using either solid board or plywood of the required thickness, perhaps one quarter or one half inch. Better still, use a brass or aluminum strip, attached with countersunk flathead screws.

This last procedure is the only suitable remedy for a damaged and mutilated plate mortise, provided there is sufficient space between the door and jamb to nail up. The new facing must be mortised for the plate, adding a definite plus to the door security, while the new wood will provide better holding power for the strike plate screws. In a metal facing strip, the strike mortise is formed by first drilling and then squaring the opening with a file.

Doubling the Jamb Thickness. In the case of the door with only a stop molding covering the door-jamb clearance, it is possible to increase the jamb thickness at the exterior portion of the jamb, by adding a strip of $1/2$-inch plywood snug against the closed door. Then stop molding can be added for weatherproofing. But the main objective, in these projects, is to remove any significant gap at the door closure.

Sticky Cylinders. A worn or grit-clogged cylinder can make any good lock ineffective. Congealed oil can cause the pin tumblers to become "frozen," that is, they stick in open position so that the cylinder core can be turned even without a key. After many years of use, the keyway can become so enlarged through wear that the pin tumbler no longer is precisely controlled by the proper key. Oil-thickened grit

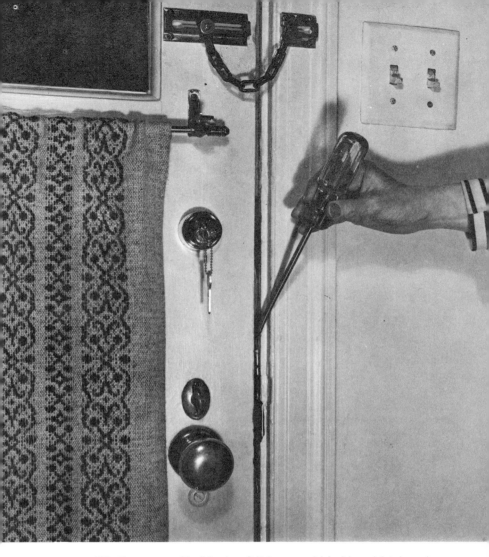

What's wrong with this door? It has an old-fashioned latch and nightbolt, and a brand new mortised deadbolt, plus an interviewer chain. But that immense gap between door and jamb makes it a cinch for a burgler to jimmy it open. The lock bolts have a "throw" or movement, of less than ½ inch, while the gap between door and jamb is over ¼ inch wide, leaving the bolt barely seated in its strike plate. Just a little shove with a jimmy or heavy screwdriver, shown here on the inside to indicate direction of the leverage, will pry out the lock bolts, and the door swings wide open.

can hold the pins in the open position so that the cylinder plug can be turned with a screwdriver, which is one reason why locksmiths caution that locks should be lubricated only with powdered graphite, never with oil.

How to Switch Cylinders. The cylinder of a mortised lock is held in place by one or two very thin and long screws which enter notched recesses in the outside surface of the cylinder to keep it from turning. These screws are on the latch side of the lock, and can be removed with a narrow-bladed screwdriver. The cylinder then is free to turn, and is removed by unscrewing it from the lock housing. The cylinder body is quite deep and the threaded section covers the entire barrel,

Changing Cylinder

The cylinder is the basic element of a door lock. There are times when the cylinder should be replaced, the key code changed, or parts thoroughly cleaned for proper functioning. In entrance door-handle locksets the cylinder is secured with retainer screws, deeply set into the door edge plate. Loose retainer screws may indicate tampering with intention to turn out the cylinder. Use narrow-blade screwdriver to turn out the screws all the way so threads can be examined.

After screws are removed, the cylinder can be taken out. Turn it counterclockwise all the way until threaded barrel clears the lock housing.

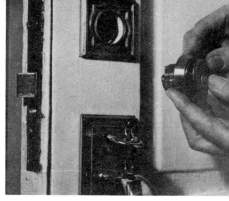

Cylinder should have a protective spring-loaded collar completely covering the cylinder rim, to prevent gripping and turning the cylinder to force it out of the lock.

|←— L ENGTH —→|

Slots on side of cylinder are aligned with the keyway position. When keyway is precisely vertical, the slots will line up with the retainer screws.

so it takes a bit of patience, but soon the cylinder is out. A local locksmith can examine it, clean the pin tumblers, and if so requested, change the key coding.

The "Pick Proof" Locks. Many homeowners display obsessive concern about the dangers of their locks being picked open. Actually, home burglaries that result from picked locks are quite rare, and those that do occur are mostly of apartment locks that are old and worn, in such poor condition that they could be opened with any kind of key, so really cannot be said to have been picked. An excessive emphasis on locks that resist pickers often causes neglect of more important and immediate lock problems. The most effective defense against lock picking is to have two cylinder locks, properly installed, plus an extra night bolt on each door.

How Locks are Picked. Picking a lock is done with long, needle-like steel tools, somewhat similar in appearance to nutpicks, with the tips bent into various shapes, of which

47

there are dozens. The picks are sold in sets for $10 to about $30, and anyone can buy them. An expert pick man can open nearly any lock about as easily as with a key, but it usually takes some time, even under the best conditions. Five-pin cylinders are easiest, but the wafer type also succumbs to an expert locksmith.

Two tools are used simultaneously, one the needle-like pick, the other a flat thin bar. This tension bar is inserted in the wider part of the keyway: the needle part is used to manipulate the pins. The bar maintains pressure in the direction the cylinder is to turn. As each pin is centered in position to clear the plug, it is held there by the continued pressure of the bar. When all five pins are aligned, the plug can turn, and thus move the latch bolt inside the lock. Sounds easy, but it's not. For anyone not an expert, picking a quality lock that is in good condition is an almost hopeless task that could require long periods of effort and usually without result. Business places, which are the targets of the more experienced and skilled burglars because of possible large amounts of cash on hand or salable goods, are justified in efforts to avoid any possibility of picked locks. The homeowner, except perhaps those known to possess valuable treasures or caches of cash, is less subject to this kind of menace. His lock selections, then, should be based on other factors.

Reinforcing Your Present Locks. Look at all your entrance doors now—not only the front door, but also the kitchen, rear, and basement doors, if you have any. Is there a mortise lock in these doors? If so, check the types. The front door lock, no doubt, is of better quality than the rest, and may be worth retaining, but how old is the cylinder? How long is it since the key code has been changed? How many keys to that cylinder have disappeared over the years? Does the cylinder work too freely, the key slip out when the latch is half turned, the cylinder itself loose in its housing so that it also turns partway with the key? All these are signs that the lock no longer is safe.

If there are any "lost" keys, they may be in the possession of persons waiting for a chance to make a clean sweep of your valuables, so a routine change of the cylinder gives you a fresh start with the assurance that comes with an entirely new set of keys. Knowledgeable homeowners keep an extra cylinder with keys on hand, to make an immediate switch whenever a key is lost or missing.

Quick Switch. Responding to the greater recognition of the cylinder as the core of door security, a method has been developed for re-keying of locks by the businessman or homeowner himself, privately, instantly. The U-Change Lock System, as it is called, allows re-keying without even removing the cylinder from the lock, and this change can be made hundreds of times without extra cost. The system is based on a special lock cylinder which replaces the standard cylinder in mortise locks, an assortment of extra keys, and a tumbler change tool. The cylinder plug is turned with its present key to a marked position and the change tool inserted into a tiny slot in the plug. The old key is removed and the new key inserted.

When the change tool is withdrawn, the tumblers are set in their new positions. The change is possible because the tool permits the pin tumblers to ride freely so that they conform to the configuration of the new key when it is inserted. Removal of the change tool leaves the tumblers locked in their new positions, to fit only the new key until another change is made by the same process.

The Halfway Key. If the key slips out when the cylinder core is turned only part of the way, you probably often leave the core half turned, thinking that the door is locked. But not so—anyone can open the lock merely by continuing the turn with a paper clip or anything else that will fit in the keyway. A cylinder in that condition needs immediate repair, or replacement. Similarly, a cylinder that turns when the key is not fully seated is faulty—the tumblers are frozen in the "open" position because of grit or congealed grease or other reason. Time for a change!

Updating Old Mortise Locks. The locks on all exterior doors—at the side, basement, or opening onto a patio—should

Remove knob and plate from mortise lock.

After removing two screws in the edge of the door, lock can be removed from mortise, sometimes with a little persuasion from a screwdriver in the knob hole.

be given the same consideration as the front entrance door. The "secondary" doors most likely have mortise locks, but these may be either the old-fashioned warded type that uses "skeleton" keys, or even if almost brand new, probably are "builder's hardware" which was selected for good looks and low cost rather than dependability.

Any mortise lock that has an ordinary spring latch — the kind that can be pushed back from the outside — should be replaced with a deadlatching lock. And all mortise locks should have a quality cylinder. Doors with glass panes, or nearby side panes, require a double cylinder deadbolt or deadlatch.

Old-fashioned mortise locks can be replaced with a more efficient one quite easily. Very likely you can find one that will just fit the door mortise or will require just a little adjustment. Take the measurement after removing the old lock. This is done by loosening the set screw on one of the knobs, withdrawing the knob stem, then turning out the two screws in the edge of the door that hold the lock into the mortise. The lock then can be withdrawn. Some locks have a deadbolt knob on the inside which also must be removed, which is done by turning out two screws that hold a metal plate to the door. If you have difficulty pulling out the lock, place a screwdriver against the knob stem opening in the lock and tap it sharply with a hammer to free the lock.

Take careful measurements of the door thickness, the width and height of the mortise (without allowance for the latch cover), and the distance of the knob hole from the edge of the door. Measurements should be precise, within eighths of an inch; otherwise you may purchase a lock that will require tedious chisel cutting when that could have been completely avoided by getting the exact size lock.

Installing Deadbolt

Mortised deadbolt supplements lockset, is particularly important installation for doors leading from basement, garage, or patio. Deadbolt may have cylinder on exterior side, or both sides.

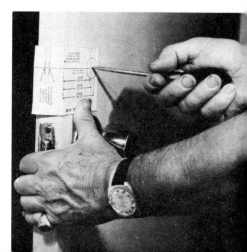

Measure from floor for convenient height, use template supplied with lock to mark for drilling on face of door and through door edge for the latch bolt. Awl or icepick makes centerpoint for drill bit.

Pilot drill with hole saw drills through the marked position, hole saw automatically is centered. Pilot drill goes through the door, to guide hole saw from both sides.

Hole saw will cut through only half thickness of the door, so drilling is finished on other side. This two-way drilling avoids splintering the veneer of flush doors.

Cutting is completed as hole saw removes core.

Edge of door is drilled at marked position for the bolt latch, which will be centered on the larger hole in the door face.

With bolt latch in place, outline of the latch plate is marked for mortising.

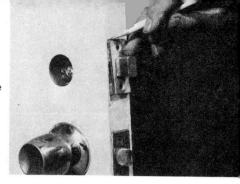

Chisel is used for carefully undercutting area to recess the latch plate. The mortise should be perfectly flat and level, so plate is flush with door edge at all sides.

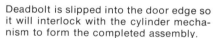

Deadbolt is slipped into the door edge so it will interlock with the cylinder mechanism to form the completed assembly.

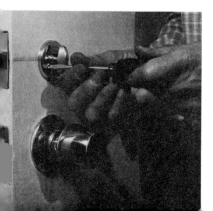

Thumbturn is fastened to interior side of door with two screws. If door has glass panes, keyed cylinders on both sides are advisable.

53

TYPES OF LOCKS

The Mortise Lock. The standard door lock, used all over the world, is generally regarded as the most effective for its purposes. It is fitted into a deep mortise cut into the edge of the door, and while there is but a very thin panel of wood remaining on each side, the lock generally is quite securely anchored.

One shortcoming of the typical mortise lock is that the spring latch, with its beveled edge for easy closing, is not an adequate safeguard. While the latch, when set by button to lock, must be opened by key, the beveled edge can be pushed back with any thin tool or a piece of plastic. This deficiency now has been overcome in some of the better mortise locks that include a deadlatch. This is still a beveled spring latch, but with a trigger mechanism that backs up the latch and makes it immovable in that manner.

The latch on mortise locks is operated by a cam at the end of the cylinder. The latter is fastened into the exterior face of the door by turning its threaded barrel into the lock housing. If this cylinder could be removed, the latch can be opened easily from outside with any tool. In some locks, the cylinder is retained by set screws driven through the latch plate on the inside edge of the door to fit into notches in the cylinder barrel. If these screws become loosened, the cylinder can be removed; and there are many opportunities when strangers or delivery persons waiting at the door could quickly loosen these screws unobserved. A protective metal cover on these retainer screws would overcome the defect — several locks do come with these protective plates. The use of non-retractable screws would not be practical in this instance.

A major item that marks lock quality is the "throw" of the latch bolt. The average movement is a half inch, and some even as little as three-eighths of an inch. When you consider that the door clearance alone takes up an eighth of an inch, at best, then the bolt "reach" becomes a critical security factor. Most quality locks, those rated as high "security" locks, have bolts that extend three-quarters of an inch, and the better ones will go a full inch.

So, in mortise locks, the selection should be based on the

Security cover by Schlage Lock Company is an auxiliary that prevents forcing or tampering with the lock mechanism. The cover can be installed over existing lock without removing it from the door. The plate is held by four case-hardened steel bolts, two inserted from each side of the door.

Improved deadlock with hardened steel bolt pin and protected cylinder collar. This Weiser Lock Company bolt has 1-inch "throw."

facts of deadlatched spring latch, protected cylinder retainers, adequate bolt movement, and overall ruggedness, with of course, quality cylinders.

Key-in-Knob Locks. Also known as cylindrical locksets, these have become the most popular in recent years, far out-selling all other types, and perhaps soon to replace the mortise lock as standard residential equipment.

The cylindrical lock is handsome in appearance, offering individuality by selection from a large assortment of rosette surface plates (commonly called "roses") in attractive designs. The lock is convenient to use, easily controlled by a push or turn button, and the single latch combines the functions of the typical spring bevel latch and a deadbolt. The way this works is that a short deadlocking pin slips behind the spring latch to prevent its being pushed back, as could be done by a piece of flexible plastic on the older type of spring latches. All locks purchased for installation on exterior doors should have this trigger latchbolting feature.

Of additional interest to the homeowner is that the cylin-

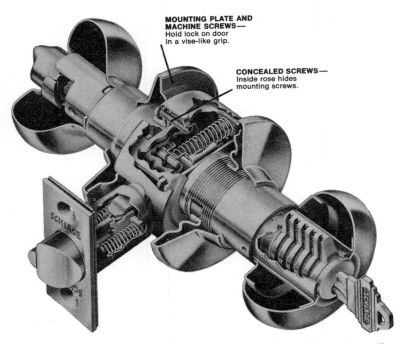

MOUNTING PLATE AND
MACHINE SCREWS—
Hold lock on door
in a vise-like grip.

CONCEALED SCREWS—
Inside rose hides
mounting screws.

Cutaway drawing shows construction details of a heavy-duty cylindrical lockset for residential installation. A wide assortment of custom hardware designs in both knob and roses is available for individual selection. Lock shown is of the Schlage "A" series which has
UL listing.

drical lock is the easiest to install, which means a substantial
saving when several locks are to be added. All that is required
is to drill two holes, one in the face of the door for the lock
case, the other into the edge of the door for the latch. Then
the strike plate is recessed into the door frame, a simple step.
The installation is aided by use of a template supplied with
the lock, and only a few special tools are needed—these can
even be borrowed without charge.

Some objections are voiced against these key-in-the-knob
locks. The most frequent is that the knobs are flimsy, and that
they possibly could be broken off, exposing the locking mechanism. Most quality locks include a backup plate on each side,
and any tampering of the knob would so distort the plate that

the latch mechanism could not be manipulated. However, tests have shown that torque applied around the knob, with a tight leather strap or a chain clamp, for example, could in some instances wrest the knob or turn the latch mechanism.

Heavy duty cylindrical locks are generally of more rugged construction, and the latch usually extends at least half an inch, and in some models up to one full inch.

Maximum security from both picking and tampering is obtained by having both a cylindrical lock and a one-inch night bolt on each exterior door. Several manufacturers recognize the value of this combination and include double service locksets in their lines. Among these are the Schlage "Double Security Entrance Lock" with both cylinders keyed alike, Sargent's "Maximum Security System," and the Russwin Locksets.

Surface-Mounted Locks. While often rejected because of bulky appearance, surface-mounted locks can provide an important extra measure of security against jimmying, particularly the vertical bolt locks which have an angled latch plate. One definite advantage is that the plate can be anchored solidly into the door frame with screws driven from two directions, while the interlocking bolt cannot be forced out of the strike plate as it can with some conventional types.

Some surface locks contain a beveled spring latch, which is locked by moving a button on the inside, and opened with a key from the exterior. Others are simply a night bolt, without cylinder. Chiefly, however, these surface locks are intended to offer an exceptionally strong deadbolt for night lockup that reinforces the regularly used lock.

Vertical Bolt Locks. Widely recognized as offering exceptional resistance to jimmying and other attacks, these locks have two vertically moving bolts that interlock with the strike plate. The lock comes with single key cylinder and turn knob on the inside, or with double cylinders for key control both inside and outside on doors with glass panes or similar vulnerability.

One model, made by Segal, has an automatic shutter guard to prevent opening the bolts if the cylinder is pulled out. The

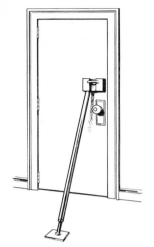

The famous Fox Police Lock features a steel bar braced against the door, making it virtually jimmyproof. When key is used in exterior cylinder, a metal block is moved aside, permitting the brace to slide upward so that the door can be opened. The bar remains fixed into a recessed floor plate, so the door swing is limited to about 40 degrees, but the brace bar can be removed from its place when full swing of the door is required.

The Fox double-bar lock, intended primarily for doors that open outward, has two steel bars that slide into hasps set into the jamb on both sides of the door. The bars are moved into position and retracted by a knob turning a worm gear in a centered gear box, or operated by a key cylinder from outside. When locked, the bars would prevent lifting the door even if the exterior hinge pins were removed. The lock can be adapted for in-swinging doors as well, using external hasp plates attached to the door frame.

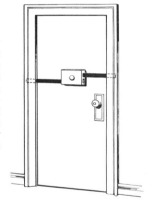

Eagle 3-Star Security model has double-bitted key cylinder and a new mechanism said to provide maximum security. Sears Roebuck lists two models, one at $8.88, another with double 8-pin cylinders for both inside and outside, at $19.66.

Fox Police Lock. In situations such as an old apartment, where the door jamb is weak or is splintered from screw holes of previous locks or where the door is a poor fit against the jamb, and where the landlord won't make repairs, the answer is the Fox Police Lock. This does not require any latch plate or other fitting on the jamb. The door—which must open inward—is held firmly by a steel bar wedged between the lock and a plate in the floor. When the cylinder is turned by the key, the bar moves aside from the stop at its end and slides through a ring as the door opens. The lock is easy to install and is strong even on an old door, because the force of attempted entry presses the lock against the door rather than straining against the screws of the latch plate as in other surface-mounted locks.

HOW TO INSTALL A LOCK

A paper or cardboard template that is furnished with most locks is essential for an accurate installation, so be careful not to accidentally discard this template. The screws provided usually are adequate, but in some situations it would be better to substitute longer or heavier screws. A case in point is when the latch strike plate is poorly anchored because the screw holes have been shifted or enlarged; with longer screws it may be possible to reach into the adjacent framing stud for a more secure fastening. When substituting screws, obtain the same diameter size, so that the head will counterset flush with the plate.

Tools Required. Mostly drills and chisels are used, varying in size according to the type of lock and the individual brand or model. Circle cutters, sometimes called hole saws, are usually needed. There are several special tools that would be helpful in making the installation and assure a correctly aligned and attractive-looking job. These may sometimes be borrowed from the lock dealer, but are not worth buying unless a large number of locks are to be installed. In that case, the considerable saving in installation cost warrants purchase of these special tools. These tools include a boring jig, about $15.00, a latch front marking tool, about $4.00, and a strike locator, which costs only 10¢. Sears Roebuck will lend customers a complete rig for installing cylindrical locksets without charge, but a deposit of $25 or $30 depending on size is required to assure prompt return in good condition.

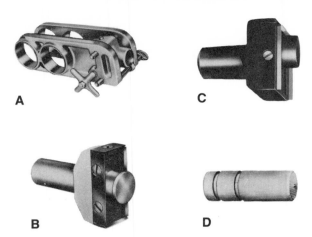

Special tools speed lock installation, assure neat result. (A) combination jig for aligning 2-⅛-inch holes in door face and ¹⁵/₁₆-inch hole in door edge; (B) latch mortiser; (C) strike mortiser; (D) dowel with center pin to mark jamb for proper strike location.

Positioning the Lock. The accepted door knob height for the average adult is 3 feet from the floor. But you may already have a lock at that position and are adding another, so the selection of position will be determined by available clearance. The additional lock should be placed above the original one, for convenience, though you may want to make an exception to this if you believe that a lock located very low, say 2 feet above the floor, will be more difficult to pick or jimmy.

There are differences of opinion about this, however, and some experts who have been consulted feel that a low lock would be even more vulnerable since the burglar would be less visible if he could sit on the floor while quietly picking the lock.

Installing a Mortise Lock. Place the template in position on the door, folded over the edge where indicated by a line. Make sure you have the correct position — the template will be marked for left-side or right-side locks, and whether it is to be placed on the inside surface or outside. Secure the template in position with pieces of masking or cellophane tape.

With an awl, carefully punch marker holes through the template for drilling, both into the door face and the edge mortise. The corners of the mortise outline should be clearly shown by the awl points.

Drill through the door face first for the cylinders and knob spindle, making the holes the sizes shown on the template. For the edge mortise you may have to drill a series of seven or eight holes with an auger or wing bit, to obtain the length needed. Draw a center line along the door edge, to guide the drill points.

A series of closely spaced ¾″ holes may be necessary for the mortise, drilled to a depth equal to the lock backset (width). Follow up with a chisel, clearing away the waste stock between the drillings, but do this carefully so as not to splinter the door facing. Smooth the mortise inside so that the lock body can be inserted. Press in the lock (but do not force it) so that the cover plate can be outlined, then undercut this outlined area just deeply enough so that the latch cover plate will fit in flush. Install the lock with the screws provided, two into the cover plate on the door edge, then the face plates and knob spindle, and finally the key cylinder which is secured with the thin retaining screws through the latch cover plate on the door edge.

To determine the position of the latch strike plate, move the lock bolt and close the door part way so the positions of the bolt and the latch can be marked lightly on the door trim. With a square, transfer the markings onto the jamb side, then place the strike plate so its openings are superimposed on the lock markings.

With a chisel, undercut the jamb to recess the plate flush, then close the door and operate the lock to confirm the fit. After making any necessary adjustments, use a ¼″ chisel to cut mortises into the jamb deep enough to receive the latch and deadbolt. Attach the strike plate with the two screws provided, after making certain that the door closes firmly against its stop molding when the latch engages.

One detail that requires some elaboration here is that of installing the cylinder. If the particular cylinder was obtained as part of the lock, it undoubtedly will be the correct size, but a replacement cylinder must be purchased to fit in relation to the thickness or "hand" of the door; otherwise it will not go in far enough, or will go too far.

The threaded cylinder is turned into the lock housing as tight as can be done by hand. Do not mar the edges with an incorrect tool; there is a special tool for this purpose used by locksmiths, but it is possible to do quite well without it. Do not use the key as a lever, as it may break off in the cylinder. If necessary insert a screwdriver into the keyway to make the final adjustment so that the keyway is vertical.

There is a useful cylinder guard with a coil-type spring at the back that takes up any extra slack and also applies tension to keep the cylinder tightly anchored. This is sold by Kramer Locksmith Supplies. The guard has a polished brass collar that covers the cylinder rim to prevent gripping and turning it.

Installing a Cylindrical Lockset. Only two holes are drilled for this type of lock. The Sears Roebuck installation tool, which you can borrow without charge, will greatly speed up the lock installation and assure a proper job. Use the template provided with lock. Place it in position at the desired height on the door (3 feet above the floor is preferred). The casing hole, which is drilled first, will be 2⅛″, 2¼″, or 2½″ diameter, depending on the lock model. This hole is centered at least 2½″ from the door jamb, to allow adequate clearance for han-

With template, mark positions for drilling door face and door edge.

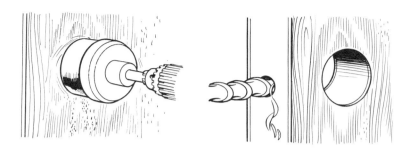

Drill face hole, 2-⅛ inches (2-inch diameter hole saw will do — enlarge hole as necessary with semi-rounded rasp), and drill edge hole ¹⁵/₁₆ inch in diameter.

dling the knob — if there is any encumbrance or other reason why the knob should be placed at greater distance from the jamb, this can be done by using a longer backset latch, which is available in standard lengths of 2⅜″, 2¾″ and 3¾″. Extension links are available up to any desired length, permitting placement of the knob and key cylinder at the center of the door if desired.

The large casing hole is drilled with a hole saw or cutter, used with a regular electric drill, as illustrated. As the cutter can enter only to a depth of about an inch, it is necessary to drill from both sides of the door — the pilot drill bit of the cutter will penetrate completely through to the other side and will align the cutter for drilling from the other side.

Next the latch hole is bored through the door edge, carefully centered with the casing hole as indicated by the template. This hole must be drilled straight and true, in right angle alignment to the door edge. The hole will be approximately ¾″ (but should be drilled to the precise dimension shown on the template, e.g. ²⁷/₃₂″) and clear through into the casing hole, a depth of 2⅜″ or more. Mark the outline of the latch face plate, and mortise for flush fit.

Now the lock can be assembled. With the button (inside) knob removed, put the lock through from the outside hole. When the casing meets the latch, it is manipulated so that the forked end of the latch mechanism slips into the clearance and engages with the casing.

Insert latch, mark outline of plate.

With chisel, cut mortise for recessing latch plate.

Install and fasten latch with bevel in correct position.

To do this, depress the latch tongue slightly to depress the latch fingers, and move the lock case farther in, to secure the latch firmly in position. Try the knob to test whether the latch responds correctly.

Next step is to turn the retainer plate and rosette on the threaded exterior side of the casing, pulling the plate up as firmly as possible so that the plate on the interior side can be locked on with two screws into the lock casing itself, thus providing a tight friction grip at both sides to prevent removal of the exterior rosette. The rosette on the interior side is held by some fastening device, a small spring or setscrew. The inside knob then is snapped into place.

Finally the strike plate position is located (this can be done most neatly early in the project by inserting a length of dowel, the diameter approximately that of the latch, and with a sharply pointed nail or pin at the center, into the latch bore. When pressed through the larger casing hole, the dowel point will mark the position of the bolt for the strike plate.) There are special mortising tools available, one made by Kwikset at about $4.00 for cutting the proper depth for full lip strip, while another tool at the same price is for mortising the latch face

Install the lock casing from exterior side, engaging the lock mechanism. Take up excess play of knob by turning on threaded shaft.

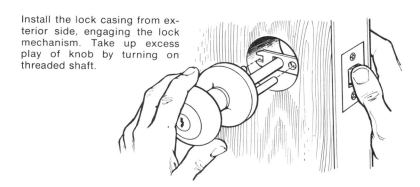

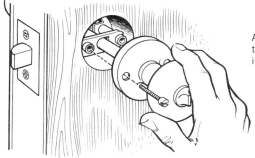

Attach retainer plate with interior screws, and install the interior knob.

Mark strike plate location.

Drill jamb for latch.

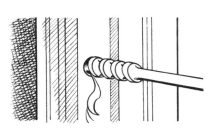

Mark the cut strike plate recess.

Fasten strike plate.

to proper depth for flush fit. If you intend to install several locks, it may pay to invest in these tools, and possibly you can recover most of the cost by reselling them.

Modernizing With the New Styles. Now it's easy to replace a worn out and old-fashioned mortise lock with one of the attractive cylinder type locksets. Special modernization kits are available that will cover the original holes in the door faces with polished brass trim plates, while a long latch plate completely hides the old mortise opening on the door edge. You are not limited to these surface plates, however, but can select from an assortment of decorative rosettes that are available with the different brands of locks, and purchase the latch cover separately. Modernization kits, which cost just a few dollars, are made by several manufacturers, including Kwikset and Rosswin. Schlage has a special G line of cylinder locksets for installation in doors which previously had other types of locks. Sears Roebuck sells a modernizing kit complete with a new exterior-grade lockset, in two sizes, catalogue No. 9P 57733, at $11.49.

How to Remove Lockset Knobs. The means by which the knobs are secured to the cylindrical lock shaft vary with the different manufacturers. There's a hint of dark secrecy about this, as very few lock brochures mention this detail. But since even the novice home burglar has learned about

This old mortise lock can be replaced by the effective and convenient cylinder lockset at right by means of a kit (this one by Kwikset) which provides fittings that cover the original openings and mortised edge. Old lock, easy to pick, had to be locked each time with a skeleton key. Lockset has thumb control that can set the door always on deadlatch.

Modernizing kit by Kwikset, including the few tools needed for the changeover. Kit includes two large plates to cover the old door face drillings and receive the new lockset case, a latch mortise cover, and new strike plate with molded edge. Most installations can be done with only a set of auger bits, but in some situations a larger hole saw will be needed.

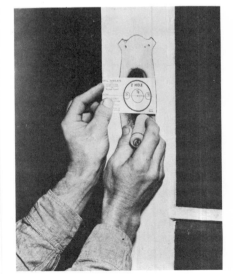

Use template to mark position of the three holes to be drilled through face of the door. The "backset" is the distance of knob stem from edge of door. The conversion usually must remain in the original backset position.

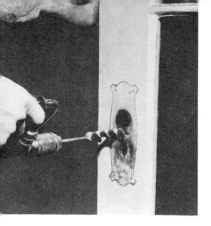

Bore two ⁷/₁₆-inch and one ⅞-inch holes at points indicated on the lock template (a single 2-inch hole drilled with a hole saw will produce the same result). Bore from both sides of the door, backing up the first penetration with a clamped board to prevent splintering the wood.

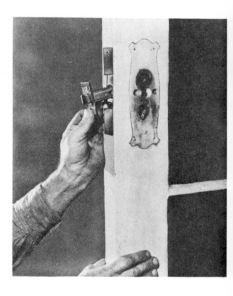

Insert the edge mortise cover in place and insert the latch all the way. If the cover plate cannot be recessed all its length, extend the original mortise as needed.

Install exterior knob and latch, then the interior knob, line up screw holes with the stems, push knob in tightly and turn in screws.

69

Remove old strike plate from jamb side of door frame.

Fasten new strike plate into position, after checking that it is in alignment with the latch bolt. Extend mortise as needed for fitting the plate neatly.

this almost as a first street-corner lesson, certainly there's no benefit in keeping the homeowner uninformed. For one thing, he should know to what degree a knob is vulnerable because of inadequate fastening, and for another, he should know how to remove the knob when necessary to make adjustments, or change the rosette or knob cylinder. Some outside knobs are held simply with set-screws, easily reached and removed. Most of the better locks have their knobs secured with concealed pins through the knob shank or spring-loaded retainer clips. The pin is very tiny, and can be punched out with the tip of a sharply pointed awl. But this cannot be done until the proper key is inserted into the cylinder and the knob

Tightening Lockset Knob

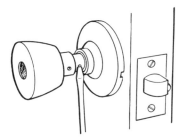

Remove inside knob by depressing spring-loaded retainer on shank with narrow-blade screwdriver. (In some makes it is necessary to force the rose inward to obtain clearance, the knob may be held with one or more screws.)

Outside knob is removed differently. First turn thumb button to lock the exterior knob.

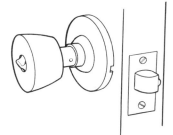

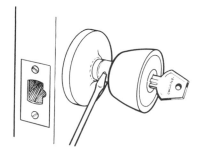

Using the key, hold the latch in retracted (open) position. Depress the knob retainer with screwdriver through the slot until the knob can be pulled off.

ob released. To replace: line up slot in ck of the knob with the fitting shape on e spindle; push knob in until it hits re- ner button. Depress the button until the ob snaps into position.

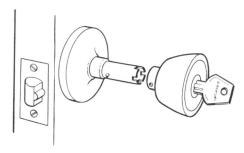

Unsnap the rose from the rose liner. Very loose knobs require tightening the rose equally on both sides, so the outside knob must be removed, as shown.

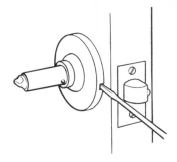

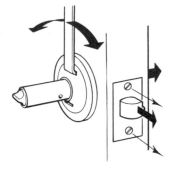

Rose liner is turned clockwise for tightening. If wrench shown is not available, a narrow-blade screwdriver can be used to turn the rose liner.

Photograph shows typical lockset with inside knob removed and the retainer slot now exposed. The rose is held in place by snapping onto spring loops of the rose liner.

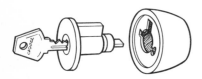

To remove cylinder from lockset handle, turn key in either direction until it can be partially extended from the cylinder plug. Then hold the knob while turning the key to the left (counterclockwise) while pulling slightly on the key until the cylinder becomes disengaged. To replace cylinder, insert key so it extends partially from the keyway, turn key to right until the cylinder body snaps into position in the knob.

turned about halfway to a certain position in which the pin can be cleared.

Installing Fox Police Lock. This surface mounted lock is easy to install compared with a mortised lock or lockset, as the accompanying illustrations show.

New Locks Meet the Challenge. Of the many innovations that have appeared in the past few years, most have been concentrated in four areas: 1. improved cylinders; 2. new keying concepts that have finally broken out of the Yale pattern which has been standard for over half a century; 3. an application of electronic controls to take the place of physical mechanisms; and 4. locks controlled by number combinations rather than by key cylinders.

Fox Police Lock is one of the easiest security devices to install. After small retainer plate is attached to floor 30 inches from the door, place the brace in it so that position of the lock can be marked. Remove lock cover by turning out two screws on surface.

Center punch marks hole for drilling door to recess the projecting base of rotating latch control pin, and for the bar from the key cylinder on the outside.

A ⅜-inch hole is drilled from inside of door; the larger cylinder hole is drilled from the outside surface.

Fasten lock to door with four screws of adequate size, preferably 1-inch No. 12, or heavier.

Operation of lock is shown with cover removed. Brace set into floor plate is wedged into blind slot of the lock.

When key is turned, the brace holder slides away from the restraining block, so end of the brace is free to move.

With lock cover on, brace is shown in normally blocked position.

With lock shifted by key, brace can slide up inside the holder loop permitting door to open part way.

The New Cylinders. Though by no means ready to be discarded, the pin tumbler lock is being challenged by various new ideas which now must face the test of time and practical experience. One of the most exciting developments is the application of magnetic force together with the standard pin tumblers. The Miracle Lock cylinder has the conventional spring-loaded pin tumblers, plus 4 individually coded magnetic tumblers — the latter corresponding to a set of magnets imbedded into the key. When the correct key is inserted, its magnets pull free the floating pins into alignment so that the cylinder plug can turn. The Miracle Lock cylinder, listed at $19.95, can be fitted as a replacement on most standard mortise locks.

A completely different approach to lock mechanism control has been taken for the Abloy cylinder, which works on the rotating disk principle. Nine of the ten disks revolve freely within the plug sleeve. The correct key aligns nine of the disks with the drop-in side bar slots so that the plug can be turned. The tenth disk is stationary, providing a bearing point

The Miracle Magnetic lock cylinder employs a combination of the standard pin tumblers with magnets. Only the correct key will release the magnet and permit the cylinder plug to turn. The key is of unique design: magnet inserts are set into dimpled recesses in positions determined by computer analysis.

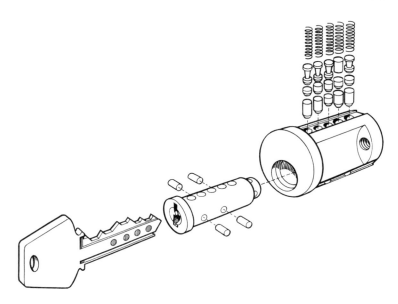

Exploded view of Miracle Lock cylinder shows placement of magnets in both the cylinder plug and in the key.

for the tip of the key as well as the cam action to rotate the sleeve. This cylinder is part of the Abloy lock, made in Finland and distributed by Intertrade Industries of Montreal, Canada.

Pushbutton Locks. The Preso-matic pushbutton combination lock has no key cylinder, thus completely eliminates the picklock hazard and is a great advance in solving the problem of lost or duplicated keys. This new lock has ten number buttons, coded into a 4-digit combination which must be pressed in correct sequence. The lock is instantly unlatched from the inside by pushing a button, allowing the door to spring open. A button, when turned as a night lock, prevents opening the door even by use of the correct number combination—an added safety feature. The locks all have deadbolts or dead-latches, and some have a spring latch that locks automatically when the door is slammed; others do not lock unless the special button is pushed. The number combination for each lock is preset at the factory, but may be changed by the home-owner at any time by inserting a new set of slides, costing $1.50. The complete locks, made by Preso-matic Lock Co., list at $20 to $30.

Dimpled Key. A completely new approach to pin tumbler arrangement and keying is the Sargent "Maximum Security System," which is not a very descriptive name, but designates a very interesting mechanical concept. Instead of the conventional cylinder plug arrangement of a row of pin tumblers which are moved into core alignment with a bitted key, the new Sargent system has three rows of pins, at right angles to each other, a total of 12 key pins in all, in scrambled sets of lengths. The key is not warded, but rather drilled for a number of shallow round recesses, on both sides and thus is reversible. The depths of the holes, determined by a computer,

Installing Pushbutton Lock

Presto! Unlock the door by pushing four of the ten buttons in a secret sequence. Door always locks when closed, but you're never locked out, because no key is needed. Preso-Matic lock shown has deadlocking latchbolt. Button at bottom clears the system so you can start over if you make a mistake in the number sequence. On the inside, one button unlocks the door at a touch. Combination can be changed by inserting a coded slide.

To install push-botton lock, template supplied with lock is placed on door at convenient height above the floor; points for drilling are marked with awl.

Edge drilling for the latchbolt is located both by template position and adjusting for thickness of the individual door. In this case metal weatherstripping along door edge had to be snipped to clear the latch bolt.

An opening 1-⅞ inches by 5-⅜ inches is required; this can be obtained by drilling three overlapping holes with a 2-inch holesaw, or by drilling a series of 1-inch holes within the required oblong area, then squaring the sides with a saber or keyhole saw.

79

Saber saw neatly trims excess stock. Angles must be squared to allow fitting the lock casing.

Sawed out stock removed showing the open section of required size. Where necessary, corners can be neatly squared with a keyhole saw.

Edge of door is drilled for 1-inch hole at marked position for latch bolt. Drill must be held to go in straight at right angles to door edge so latch bolt will function smoothly without binding.

Dowel of 1-inch diameter, with pin at center, is placed in latch boring to locate the strike position on door jamb. Dowel is pressed outward from the open section.

Lock body is set into the opening from exterior surface, locked into position with cover plate on the inside, at the same time engaging the latch which is inserted from the edge.

Lock completely assembled, showing the two inside buttons. One releases the latch instantly for normal exit; the other button sets the night latch so that the lock cannot be opened from outside even with the right combination.

control the repositioning of the pins so that they become aligned to clear passage for the plug.

The Sargent company says that the keys cannot be duplicated on key cutting machines now in use, and that all duplicate keys must be cut on special equipment at the plant, and are supplied only on authenticated orders of owners. There are said to be a total of 24,500 key changes available within each of seven complete master keying systems available.

Electronic Security. A new vista in entrance door control has been opened with the development of electronic locks. One of these is the Cypher Lock, which presently is used for security control in industrial plants and offices, but is expected eventually to be adapted to residential use also. The unit consists of 10 numbered buttons. As with the Preso-matic lock, a combination of four of these buttons pressed in the correct sequence will activate an electric door opener. The unit and door opener operate on low-voltage current off a 6-volt battery. An interesting feature of the Cypher lock is a "time penalty." If an incorrect or out-of-sequence button is pressed, the lock will block any further efforts to open the door (even with the correct combination) for a specified time period. The time penalty may vary from one to 10 seconds, and has the purpose of preventing repeated attempts to discover the combination by trial and error. The Cypher lock is a product of Sargent & Greenleaf, Inc.

6

Looking at
the Basement

The basement seems to be at the bottom of the list when the homeowner considers security arrangements. This area deserves, instead, the prime consideration, since it is most likely the weakest link in your defenses. It is farthest from the sleeping quarters, the doors and windows usually are of flimsiest quality, deep window wells shield an intruder from view and muffle sounds of forcing an entry, and because the basement provides hiding places, you may be unaware that an intruder has broken in until it is too late to summon help.

Cellar doors offer little resistance to entry. Almost all have small glass panes intended to provide some daylight to the otherwise poorly lighted basement area. A broken glass gives quick access to the interior throw bolt. Even without vulnerable glass panes, the basement door is a pushover. Locks are surprisingly inadequate; nearly always they are of the common mortise or rim type that can be opened with a skeleton key — or just a vigorous push!

Additionally, the hinged sash windows, so tiny that they

seem an unlikely means of entry, are an open passageway to burglars who don't mind crawling through such openings. Often the prowlers are accompanied by a young boy who climbs or creeps through just such neglected spaces, then opens the doors from the inside. Any opening that is even a foot wide should be regarded as a possible invasion point.

Alarm Link Important. Many homes that have fairly complete alarm installations somehow fail to include the basement entryways into that system. Just because the basement area is out of your view doesn't make it the less visible — and inviting — to an intended intruder. Look over the entire cellar area again with the objective of linking every door, window and other possible entryway into your burglar alarm system if you already have one, or your plans for a system that you are starting to install. One big advantage here is that the alarm circuit wiring will be much easier because there won't be any real problem about snaking the wires or fear of damaging the room decorations.

Basement Exterior Door. If the door has glass panes and you want to retain them, put up a heavy wire guard over the entire glassed area, firmly anchored so that it can't be pried off easily, or glaze with additional panes of glass having embedded wire mesh. The double glass will be more difficult to break, and the wire mesh will, mostly, serve as a deterrent. Better still, use sheets of transparent plastic, either Lucite or General Electric's new product, LEXAN, which is virtually smashproof and has been designed primarily for burglarproofing.

Some doors are glazed with putty around the glass, others have wood molding as retainer strips. So be careful not to fool *yourself* by glazing the door in such a way that the panes can be easily removed from the outside, simply by prying off the wood molding strips. Here's how to overcome this defect: if the door is so built that the retainer molding is on the outside, drive a series of glazier's points all around the new glass or plastic inserts before puttying or nailing on the molding. The points are difficult to locate and remove, and will effectively reinforce the panes.

Plastic sheeting is a practical replacement for glass in vulnerable door panes and windows. Lexan, developed by General Electric cannot be shattered even by hammer blows. It is cut with any woodworking saw. Allow ⅛ inch clearance on all sides of sash for expansion. Leave the protective paper on until installation is completed.

Clean sash rabbet and apply elastic putty or special tape. Fold back protective paper and press the panel uniformly into the bed of putty. Apply putty as with any pane, and remove paper. Clean any smudges with soap and water; a razor blade will scratch the plastic.

Better than doubling up the glass is covering completely both sides of the door with ¼-inch plywood or hardboard panels. This will reinforce the door itself, adding to its thickness while blocking the vulnerable glass. The light that is lost thereby can be supplied instead by installing one or more fluorescent fixtures. The 2/40s (two lamps of 40 watts each) give excellent illumination.

The plywood can be laminated to the door without clamps, using contact cement. Lift off the door by pulling the hinge pins, place it conveniently across sawhorses or other support, and remove the lock and knobs, but leave the half hinge in place. Cut the panel for each surface to precise fit, notching around the hinge leaf, and also marking the place of the spring-lock knobs on the surface of the panels. If the mullions around the glass panes are not flush, plane them down as necessary, or even remove the moldings. Now coat the door and the reverse side of the plywood panel with contact cement spread with a piece of scrap hardboard, allow the cement to dry about 20 minutes. Test for dryness with a piece of kraft paper — if the paper doesn't stick, the cement is sufficiently dry.

A large sheet of kraft paper placed over the door helps align the plywood with the door edges before the cemented surfaces touch. Place the panel on the paper, and carefully square the edges all around. Put a couple of brads part way at one end to hold the panel in place, then lift the other end and fold back the paper underneath about a third of the way. Let the panel set down in place and press firmly. Now you can remove the temporary brads, lift the panel enough to pull out the paper, and ease down the rest of the panel, which should be in complete alignment with the edge all around. Press firmly with your hands, then tap the panel with a wood block and hammer to be sure it is well bonded.

Do the same on the other side, then replace the door on its hinges. It will be necessary to adjust the door stop moldings so there will be sufficient clearance. After the door is back on its hinges, the stop position can be determined and the molding nailed back in place. Usually there is no need to adjust the door threshold. The spring lock knobs can be replaced as before, if the same lock is used, or a new deadbolt lock set into the mortise.

Better Lock Needed. The basement door warrants better locks than have been assigned there in previous times. A mortise lock with half inch latch plus an extra deadbolt make a solid combination. One of the more dependable locks for such unguarded places as the basement is the Segal or Eagle vertical interlocking bolt. Even greater security is obtained with the Fox Police Lock, or with the Kramer Double-Bar Security Lock, both fully described in Chapter 5. The latter is best used without an exterior keyed cylinder, and usually one is not necessary for the basement entrance. The objective for this door is a maximum protective barrier against forced entry, while retaining the means of rapid exit in an emergency.

Cellar Hatch Doors. Wood hatch doors do not offer adequate protection, no matter how carefully locked, since the wood planks can be pried up quite easily with a crowbar. Then the enclosed stairwell provides full concealment for a prowler determined to break his way through the inner door. Only steel hatch doors, securely bolted into the concrete base, provide a sufficiently strong barrier. They should best be barred from the inside. Where an outside padlock is used on hatch doors, the hasps should be of the shielded type to prevent the lock being forced apart.

Protected Cellar Windows. The extent of window protection for the basement, and the type of guards used, will de-

Apartment windows or those on lower floors of homes can be fitted with ornamental grilles, made to any size, for burglary protection. The grilles may be permanently fastened to the window frame, or one side hinged while the window is fitted with a quick release latch for emergency exit. Padlocking is permissible only where other means of exit are available.

pend largely on whether the windows are necessary for emergency escape routes. Where there are several windows, one can be selected as an emergency exit, while the others are fully blocked off. The window reserved for exit should be one that is least susceptible to attack by a prowler; that is, its location is easily seen from outside, under ample lighting at all times, and difficult to reach for one reason or another. This exit window also can be protected, but with a mesh guard that is hinged and has an easily opened catch on the inside. The other windows then can be more securely protected with fixed installations.

Flat steel bars, ½-inch thick and about 1 inch wide, can be bolted to the window frame, spaced no more than 8 inches apart either horizontally or vertically across the window. A less expensive installation can be made with ¾-inch galvanized pipe, threaded at both ends and turned into steel pipe flanges which are attached to the frame through the flange screw holes.

Wire mesh makes a satisfactory barrier if the metal is of No. 8 or No. 10 steel wire, welded into a ⅜″ or ½″ round frame. These wire guards are available with various types of burglar-proof hinges and closing fasteners to meet different installa-

Iron bars fitted to basement window frames protect one of the most vulnerable, and difficult to watch, means of entry.

tion conditions, from local iron works or the Kentucky Metal Products Co.

Folding gates also offer a means for window security. The gates ride in top and bottom channel tracks, have riveted scissor bars with spaces of not over 6 inches when the gate is closed. Padlock fittings or cylinder locks, or a secret release catch, may be used to provide an exit in emergency. These gates are also available from Kentucky Metal Products Company.

Window Well Guards. Dependable protection of the basement is provided by iron gratings or heavy wire mesh guards installed outside over the window areaway. Essential requirements are that the guards or gratings be of adequate strength, and securely attached to concrete areaway walls. A grating of 1/2-inch by 1/2-inch bars, with welded joints, is most effective where the window is almost entirely below grade level. The grating should be installed so that the corner segments are deeply set into concrete. If the window is used occasionally for passing lumber into the basement, or similar purposes, the grating can be arranged to pivot upward at the end closest to the house wall, the front end held with a padlock accessible only from underneath.

Heavy wire mesh guarding basement window also serves as protective surface against accidental falls into the areaway. Grating consists of galvanized No. 8 wire in frame of 5/8 inch steel stock which is securely anchored to the areaway curbing.

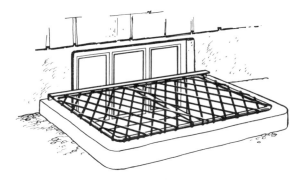

Window well gratings of No. 8 steel wire with 2-inch mesh are also practical, particularly in situations where the part of the window rises some length above ground level. The grating then can be sloped from the wall to the front edge of the areaway, with the sides similarly enclosed. These gratings, built with a heavy metal frame, are securely bolted to the areaway concrete.

For semi-circular metal wells, a new type of cover is available that offers at least some additional barrier to cellar entrance. This is a plastic bubble, made of clear Lucite, that fits over the entire areaway and covers the above-surface section of the window, with sides of ¾-inch exterior plywood. Flanges along the top and sides permit counterflashing, and the frames may be attached to the house wall with masonry anchors. The plastic bubbles range in price from $16 to $30 and more, depending on size and type, and may be ordered from the Dilworth Manufacturing Company. The translucent plastic does not block the light at the windows, and the covers serve also to shed heavy rains that in some homes result in flooding.

Full window guard and areaway grating of 2-inch wire mesh provides for maximum daylight in interior while serving as a dependable barrier to intrusion. The window guards are available in stock sizes or to order from Kentucky Metal Products Co. Special lugs serve as hinges for gratings that need to be lifted occasionally for passing lumber and similar materials into the basement through the window.

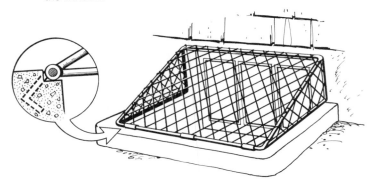

Looking at Skylights. In many urban homes, particularly town houses or "row" houses, some rooms have skylights that can be reached over the roof from an adjacent house. Any skylight is easily forced open, and it's no problem for the prowler to drop down to the floor.

Skylights have always been difficult to safeguard. One step is to see that the skylight glass is the heavy, mesh-reinforced type. Better still, replace the glass with clear Lucite plastic, at least ¼″ thick, which cannot be smashed easily, or the new LEXAN plastic by General Electric.

A more effective measure is installation of wire mesh guards, either above the skylight glass, or underneath across the ceiling opening. Guards placed above the skylight should allow a clearance of several inches above the glass. For installation in the area underneath, special supports and fittings are available.

More important, include the skylight in your burglar alarm installation. One way is with aluminum foil circuit on the inside surface of the glass. This application will most likely be very difficult, however. A simpler, yet fully effective measure, is the use of trap wires strung across the ceiling opening. There also is a special uninsulated wire, No. 24 gauge, listed by Alarm Devices Mfg. Company and made especially for skylight alarm installations.

Interior Connecting Door. However carefully you've secured your basement area against entry by prowlers, don't ease up on the last but still very important barrier. That is the interior door, leading from the basement to the living area of the house, which stands as a secondary guardian of your family's safety. With all you're planning now for the basement security measures, it's important also to reinforce this door. This is particularly important in homes with an attached garage, or garage space under part of the house, from which the entry to the house is directly through this basement passage door. Once a burglar gets into the garage or basement, he can work on that door at the head of the stairs in complete concealment, using every trick in the book to break in without fear of detection. If he's sure there's nobody at home, he can use any force at all — even an ax if he finds one in the basement or garage — to crash the door, and never worry about all the noise.

A solid deadbolt latch is advisable — one with a key cylinder on the approach side if used for entry from the garage. A chain with lock attachment also is worthwhile, even though it may involve some inconvenience, though this can be minimized by reserving the chain for times when the entire family is at home. Most important is some signal to let the occupants know each time the basement door is opened — this can be a "make and break" switch which rings just once for each opening of the door, at all times of the day and evening, with a supplementary switch that throws that door detector onto the full alarm system during the night. The circuit installation remains the same as for the rest of the system, but with a 3-contact lock switch at the basement interior door so that the detector can be switched to a separate bell when desired.

7

Help for Apartment Dwellers

Apartment residents face extremely difficult and trying problems in countering constant thefts, threats to personal safety, and invasion of privacy. It would seem that apartments are easier to protect than a private home—there's just one entrance door, only a few windows (which are inaccessible except for those at ground-floor level or facing the fire escape), neighbors on all sides, and a staff of building employees at hand. Nevertheless, the amazing frequency of apartment thefts and physical assaults at all times of day and night proves that safety is not to be taken for granted. The doors and fire escape windows of the apartments are exceptionally vulnerable, while danger lurks constantly in the elevators and corridors, at the mailbox alcove, in the laundry room, the basement garage, and in an unattended lobby at night.

These dangers cannot be overcome with a simple latch or two. Rather, a whole range of defense concepts must be put into effect, aimed first at making the apartment as impregnable as possible, then at strengthening the security of the building itself against multiple dangers. The one goes with

93

the other—you're not likely to achieve serene occupancy of
your apartment in a building where doubtful characters can
come and go without question, where an unlocked or carelessly supervised lobby provides convenient hiding places
for waiting muggers, where dimly lit corridors are an invitation to attacks, where other tenants may in fact be the burglars
you are trying to avoid, and where passkeys are under haphazard supervision.

The inescapable fact of apartment living is that the buildings are amazingly easy for prowlers to enter and to operate
within at will. Do not be lulled into complete confidence because the building has doormen on 24-hour duty or all-night
security patrols. These turn out to be less than reliable, at
best, while service entrances are locked only late at night, the
garage attendants are frequently absent making car deliveries,
or busy with car washing, the stairways offer unchallenged
hiding places, and the quiet corridors provide excellent cover
for burglars picking a lock or forcing a door.

Tightening of the building security cannot be accomplished
by the individual tenant. It can result only from the combined
urging of many tenants on the building owners or management, together with the responsible participation of all the
tenants in a security program. Management can cooperate by
installing better locks on all entrances, keeping unsupervised
doors locked at all times, paying closer attention to screening
of employees, providing better lighting at the entrance and
courtyards, in the lobby and corridors, keeping the doorbell
annunciators in working order, assuring more careful control
of the master keys, and installing window guards where requested on lower floor apartments.

All these efforts can be negated, however, if a single tenant
fails to cooperate, carelessly admits anyone who buzzes the
downstairs buttons, blocks a service door on leaving so it
won't lock because he's forgotten the key and won't bother
to go back for it, or invites crowds of barely recognized ac-
quaintances to parties.

Armed Enclaves as Refuge? The National Commission on
the Causes and Prevention of Violence in its report to the
President forecast the extreme methods that would be used

to assure privacy and security in high-income apartment buildings, describing the measures as resulting in "fortified cells." Some new housing developments, have, in fact, confirmed this grim prediction, placing guards at gatehouses of residence compounds like sentries at a military camp, barring entry to any visitor until clearance is obtained.

The grounds are monitored by elaborate electronic security devices such as closed-circuit television, proximity alarms, and radar detectors. While these provisions originated at very expensive apartment buildings, similar measures are now included in a number of modest garden-type developments. Carrying the picture a bit further, apartment complexes have cropped up which serve as secluded enclaves, comprising self-contained exclusive communities with clubhouse, swimming pool, kindergarten, shopping—all making for a stay-at-home existence under the protective eye of a private guard patrol. Will these places be indeed immune from criminal incidents? That remains to be seen, but experience indicates that it is doubtful. In any event, such sheltered surroundings are not available to all, certainly not to those who must work and live within, or close to, the city limits, nor would everyone choose such a closely guarded home environment. The average citizen, perhaps, is willing to accept and cope with the risks that come with more ordinary living circumstances.

Apartment Security. Apartment security calls for carefully selected hardware on doors and windows, competently installed, as a first step. Additionally, unceasing attention to details is essential, including double-locking of the door even for brief periods of absence, strict control of the apartment keys, and proper protection of the windows. The fire escape has been put there to protect you in the event of fire, but at the same time it is an ideal means from a burglar's standpoint to force his way into an apartment through the window. Remember, too, that terrace balconies on high-up floors can be scaled from the roof, or from an adjacent terrace.

How Many Locks? Pointing up the absorbing attention to the subject of door locks by so many urban apartment residents is the yarn about a woman who had an eighth lock installed on her apartment door because the previous seven had been picked open so often. But just hours after the new lock was installed, she rushed back to the locksmith to complain that the new one also had been opened. The locksmith, the story goes, suggested that if she locked only four of the locks it would prevent the burglar from entering since his efforts would result in locking as many locks as he unlocked.

The scheme must have frustrated burglars no end, since the woman happily reported that from that day on, all the locks she had locked were open when she returned home, but all the locks that had been left open were then found locked.

This unique approach is not presented as a recommendation, but the story does point up the importance of adequate door locks as the primary apartment defense barrier.

The lock that you find on the apartment door when you move in should be looked at with a suspicious eye. Very likely, the lock wasn't of a very dependable quality to begin with, its cylinder keyway most likely so worn or gritty that almost any key will turn it, and the lock probably is the spring-latch type without a self-bolting feature. Then, you've no idea as to the key situation; the chances are that an unknown number of the keys are floating around in questionable hands. Assurances of the super as to the master key control need not be accepted at face value. He very likely is honest and careful, but you can't keep tabs on his staff changes.

If you are required by your lease to retain the original lock, you certainly should insist on having the cylinder changed, preferably for a brand new one. This still leaves the master key susceptibility, but at least the responsibility has been narrowed down. An interesting development, applying to new apartments, is the Russwin Construction Key System. This is a door lockset for which workmen use one set of keys during the construction work. After completion of the building, the first occupant receives a different set of keys to the same lock. When the new resident uses his keys for the first time, a change takes place in the cylinder pins that makes the former construction keys inoperable and useless. This is accomplished by an ingenious split pin device in the lock cylinder, and the system may end a long-standing problem concerning possession of keys to new apartments by unknown persons who had worked on construction.

Value of a Second Lock. A second lock on the door is a worthwhile investment that practically doubles your door security. This addition which is almost essential if you're taking over a previously occupied apartment, helps make the door that much safer against jimmying as two widely separated bolts would have to be pried apart from their plates. Another advantage is that you can have two different types of locks: one for example, the conventional mortised bolt, the other a solidly mounted surface type with pickproof cylinder and double-arm bolts. Examples of suitable apartment locks are the Kwikset, Segal, Russwin, Schlage, Sargent and Abloy. These, and others that are also recommended, are described more fully in the separate chapter on Locks.

A prime requirement is that at least one of the locks be of self-latching type, so that you know the lock is set just by closing the door, even before turning any deadbolt knob on the inside, or taking the double turn with a key from the outside. Then, when you come home with arms loaded with packages, just let the door slam and you know it's locked — so important to keep an intruder from following you inside. The system works, too, when you leave the apartment hurriedly for just a brief period — to get the mail or take your laundry out of the machine — and there's the tendency to skip throwing the door bolt with your key as you leave. The self-locking provision thus provides reassuring security, keeping the door secured at all times.

You can, and should, go a step further in your lock selections. Of your two locks, one should have a deadlock bolt throw of at least 3/4", with a tamper-protected cylinder.

Important as the choice of locks surely is, the condition of the door itself, and proper lock installation, are two additional factors that weigh heavily in determining the strength of your defenses.

Testing the Door Condition. Perhaps you've never given close attention to the door before, but this must not be overlooked. There isn't much that can be done to buck up a sagging, battered door, with cracked stiles and warped panels, short of replacement. Many apartments have metal doors with wood filler cores. Test the center panels, since some are made of extremely thin sheeting that can be easily separated with a kick from their retainer grooves, enough for a hand to enter and shift a deadlock bolt or remove a night chain.

If the metal does not meet your test, and you can't get the door replaced, it can be reinforced with an extra panel of 1/4″ plywood on the inside surface. This is not too difficult to do, and the door need not be taken down. Permission of the landlord usually is required, and perhaps also a check of local fire laws to determine whether a plywood facing would be a violation. A sheet of steel or aluminum may be used instead to good effect, attached with sheet metal screws.

Details on laminating a plywood or hardboard panel to a wood door are given in the chapter on basements. For attaching a panel to a metal door, use self-tapping screws in pre-drilled holes, or roundhead wood screws if there is a wood core in the door frame. Locate the screws at regular spacings to form a neat pattern.

Many apartment doors bear the scars of repeated attempts at entry which left the door edges deformed and weakened. This condition should be called to the attention of the superintendent with a request for a reinforced edging and new stop mouldings that cover any space gap between the door edge and the frame. A length of 3/4″ x 3/4″ steel angle iron, attached along the door edges, will add immeasurably to the apartment's security. It must be secured with non-removable screws, whose heads engage a screwdriver in only one direction. This won't be very attractive, but it will surely add to your safety and peace of mind.

Transom Treatment. A transom may appear too small for anyone to crawl through, but it can allow a prowler to reach down and unlock your door from the inside. The best thing to do with an over-the-door transom is to block it off with a solid panel, or if you wish to retain its light value, with steel bars or a heavy wire mesh guard framed with heavy tubular steel. If these measures prevent use of the transom for necessary ventilation, the alternative is to install a mercury-type alarm switch, described in Chapter 11, which sounds an alarm when the transom door is tilted downward. This installation requires wire connections to a bell powered by either batteries or a transformer plugged into an electric receptacle.

In some cities, regulations require that all apartments be provided with a door interviewer. If this is lacking in your apartment, you can install one quite easily with an electric drill as described in Chapter 6.

Door Chains Useful. Although you may have satisfactory locks and one-way interviewer glass to identify callers, the additional protection of a door chain is too good to pass up. Should an intruder pick the lock, or use a wandering key, the chain guard becomes an additional barrier to entry. The type

Chain with lock can be attached from outside when leaving the apartment.

Massive door lock made by 3M features a tamper alarm and inter-viewer chain. The alarm sounds when an attempt is made to force entrance or tamper with the lock cylinder. The alarm is powered by flashlight batteries, requires no wiring. Close up view of surface-mounted lock shows easy-to-use pivoted bolt control bar. This lock has extra-heavy bolt that moves almost a full inch into the solidly mounted latch plate.

that can be locked from the outside when you leave the apart-ment serves the double purpose of an extra lock; while an attempt to open the door while you are at home will give you additional notice and time to telephone for help. Some door alarms have integral guard chains, sounding off at any pres-sure put upon the chain when an attempt is made to force open the door.

Among the alarm type of chain guards are a model made by the Slaymaker Lock Company, listed at $6.95; the Alarm Lock made by Leigh Products and priced at $5.95, and the Alarm Guard, product of Wessel Hardware Corporation, priced at $8.95. Also, the 3M Alarm Lock includes a heavy door chain.

Bolt and burglar-alarm combination is useful for apartment doors. Alarm is set when bolt is locked into the strike plate. Pressure on the door against the bolt sets alarm ringing. This Stanley self-contained unit is operated by readily available C battery. Battery must be checked regularly to insure protection.

Sliding windows, patio doors, and the like can be secured with a wraparound chain that holds against any pressure. However, it offers no protection against entry by breaking the glass.

These use flashlight or transistor batteries. Once the alarm has been set off, it can be silenced only by someone inside the apartment.

An easy-to-apply chain guard, made by the Ajax Corporation, is in the form of a loop that slips over the door knob. The chain is of welded steel links, brass plated, and supplied with 2-inch-long screws for solid mounting of the hasp on the door frame. No installation is required on the door itself. Standard door chains are made by Safe Hardware Company, which also produces a type called Safeguard, with a unique maze-like retainer plate that balks attempts at release of the chain from outside the door.

Sash lock prevents opening window even if glass is broken. Mounted on top of bottom sash, bolt inserts in top sash.

Sash lock slides on plate which is screwed onto top of lower sash.

Drill and screwdriver are all the tools needed to install sash lock. Place lock in position and mark for bolt hole with window tightly shut. Drill another hole vertically above at a distance that will permit the window to be locked open for ventilation, but will prevent entry.

Window Latches and Alarms. Several types of latches permit setting the window sash at various open positions, preventing further movement. Window openings should be limited to six inches. One standard protective measure is to drill a small hole in a top corner of the lower sash, and a series of matching holes in the upper sash in such a way that a nail through the holes locks both sashes in place. The nails cannot be reached from the outside when the window is open just a few inches.

There is always the danger that the glass will be smashed by a prowler standing on the fire escape. This is more likely to occur during the daytime when tenants are away at work and the street sounds cover up the noise of breaking glass. Means of window protection are limited, but you have a choice of several effective methods including metal guards, alarm detectors, and window shutters. The fire escape ladder in your building ends at least one story above the street level, approximately 12 feet, so it would seem unlikely that anyone could get up those stairs to your apartment window. This obstacle does not stop any burglar — there are ways to pull down the balanced retracting ladder so it can be reached from

Spring snap bolt for latching windows can be used on both double hung and sliding types.

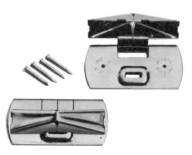

Just flip the lid and the window is secure, with this type of basic latch, which is one of the apartment security devices made by Ajax Hardware Company.

A variation of the basic snap-type window latch, also an Ajax product. This one is especially designed for sliding windows.

the street, or a simpler method of taking the elevator in the basement up to the roof, then starting down from there. Easy! On the way, a burglar can pick the most likely victims, often making a clean sweep from one apartment to another on the way down.

These windows, and all windows that can be reached from the ground floor level, need special attention. But how can those windows be protected? Keeping them shut and latched is not always practical in an apartment which has few enough windows for ventilation. Air conditioners luckily offer a great help. While they may not be placed on windows facing the fire escape, because that would interfere with emergency exits, the air conditioners make opening of windows for ventilation less necessary.

Metal Window Guards. Permanent fixed bars or a metal guard at some windows may be forbidden by the fire laws in your community, and for good reason; blocking those windows could cut off escape in the event of fire or other emergency, and also hinders firefighters from getting at the blaze quickly. However, there are some types of guards which do not present this hazard, and would be a legal installation. One type consists of folding gates with a latch that is easily opened

Typical installation of ornamental screening at lower-floor windows that are highly vulnerable to entry by intruders. The screening also serves to avoid glass breakage from vandalism.

from the inside but is inaccessible from the outside because of a wide metal plate. The Protect-A-Gard gate, manufactured by Windor Security Systems, folds out of sight into a sort of pocket when closed. The gate is quite massive and unpleasant looking for home use, however, and would be selected only where the situation is so susceptible to danger that no other means would be satisfactory.

Another type of guard that could be used at both fire escape and ground floor windows is manufactured by National Guard Products, Inc. Made of heavy gauge steel angles and strips, the guards come in various stock height and width sizes, ad-

Four of the many designs in ornamental treillage that can be used for protective screening of lower-floor windows and other vulnerable openings.

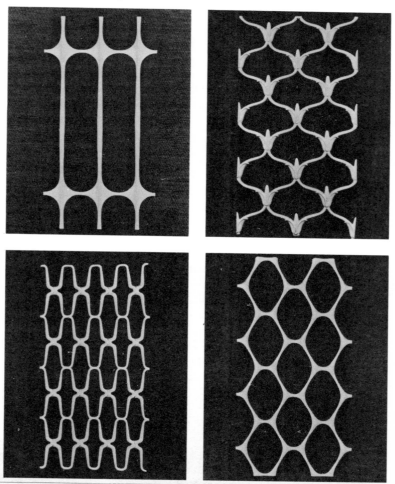

justable for precise fit into the window opening. The guards are designed to be attached to the window frame sides with screws. There is also a way to install the guard so that it can be removed rapidly from the inside should that ever become necessary in an emergency.

This plan uses the keyhole slot principle—a larger hole is drilled immediately below and contiguous with each attachment screw hole. These holes are large enough to permit the head of the screw to go through. The screws are turned in just tight enough to permit raising the guard. When the guard is lifted only a fraction of an inch, the adjustable sections are pulled inward just enough to clear the screw heads, and the entire guard can be taken out, without use of any tools.

The screw heads cover most of the keyhole slots so they are not visible and attempts to remove the guard are not likely, while the mere presence of the guard would act to deter any prospective burglar. Of course, for permanent installations at ground floor windows, the guards could be more solid and offer greater protection. Prices for stock size guards range from approximately $7 to $10 each. The guards need not cover the entire window area, only to a sufficient height that they cannot be bypassed.

Window Burglar Alarms. The best solution to the problem of window security, and to apartment protection generally, is a quality alarm system. Alarm units are flexible and there are components to meet every requirement. Self-contained, one-station alarm units are of simple construction, inexpensive, powered by flashlight batteries, and easy to install. Their chief purpose is to alert an occupant to an intrusion, and that is certainly a valuable service. There are several objections to these low-priced units, chief among them being insufficient sound volume (despite the claims of advertisements), the fact that the alarm can be turned off almost instantly by an intruder just by pushing a lever, and usually shoddy construction, which makes these units not quite reliable.

Much better types are available with key lock "on-off" controls, time delay, and test indicators. A more advanced and reliable system that requires circuit wiring is based on magnetic detectors; one element is attached to the window frame, the other to the sash. The sash can be raised a few inches for ventilation, but going beyond a predetermined point will set off a loud alarm bell located in a separate part of the apartment or in the hallway. The most effective system for window security involves metal foil strips cemented to the glass as part of the alarm circuit. Details on these and other alarm systems, with installation and operating instructions, are given in Chapters 9 to 12.

Ultrasonic Alarm. A dramatic new approach to security in apartments is an electronic intruder alarm. This is an ultrasonic device that senses motion within a specified cone-shaped area of inaudible high-frequency sound waves, and responds by both flooding the room with light and sounding a shrill alarm. The unit is completely self-contained in a wood or metal case that looks like a small radio, is installed merely by plugging into a nearby electric receptacle, and is particularly suitable for apartment protection. Prices range from $100 to $150. The effectiveness of the unit is increased by supplementary wiring and alarm bell installations.

The unit covers a cone-shaped floor area of 100 and more square feet (depending on the model or type) as shown in the illustration. A lamp cord is plugged into an outlet of the alarm cabinet. When any person moves into the area of the ultrasonic beam, the unit first switches on the light—which can have a startling effect on an intruder. After 15 seconds delay, the alarm siren goes on. The time delay has two purposes: it adds a surprise effect to scare off an intruder, but also gives the occupant an opportunity to shut off the alarm if the switch has been tripped inadvertently or it becomes necessary for the occupant to cross the controlled area.

The housing of the device has terminals for connecting a remote bell or horn with low-voltage wires, and the alarm siren itself can be silenced. Thus, when the alarm goes off, the intruder would not be able to trace the source to the alarm box, since that is not distinguishable by appearance. If desired, the remote bell may be located outside the apartment, possibly in the hallway, if that would bring help.

If you attach a remote bell to the unit, the bell should operate on its own 6-volt lantern-type battery, and have a constant ringing drop relay (described in Chapter 10) on the line. Thus, even if the intruder quickly finds the alarm box and unplugs it, the remote bell will continue to sound, on its own power source. If this bell is placed high on a hallway wall, or in an adjacent apartment, or some other location that can't be reached easily, it will provide maximum effectiveness in scaring off the intruder or bringing previously arranged assistance, possibly from the building staff or neighboring tenants, or notification to the police.

Typical of these ultrasonic alarm units are: 3M Intruder Alarm, by Minnesota Mining and Manufacturing Co., price $125.

Sears Roebuck Ultrasonic alarm, with separate horn, No. 57P 7225, price $99.50 (may be used as central alarm station with window and door detectors).

Theft Prevention Company, Ultrasonic Alarm, Model P-101, price $109.00, Model P-106, price $149.50.

Delta Products, Inc. Deltalert, $69.95, plus horn, $24.95.

Bourns Security Systems, Inc. ultrasonic alarm.

Emerson Electric Co. (Rittenhouse) Alert-Alarm unit.

Shutters Coming Back. One additional possibility for window protection is the use of hinged louver shutters, installed either on the outside or inside window frames. Such shutters are widely used in France, Italy, and other European countries for both privacy and security. The shutters are conveniently opened and closed from the inside, and the louvers are adjustable to provide plenty of light and air while barring prying eyes.

Wood shutters are the most frequently used. They can be purchased ready-made in pairs at any lumber yard, and can be planed down for precise fit. Aluminum and steel shutters of this type are also available. For outside installation, the shutters are hung on pivot hinges attached to the house wall or to the window frame. Inside shutters, easier to install and manipulate, are quite popular for their decorative effect. A sturdy but easy-to-use inside latch can be attached to lock the closed shutters. A flat steel bar placed across the shutters in retainer brackets will provide additional security.

Elevator Manners. Apartment residents coming home late in the evening must pass two hurdles before finally reaching their apartments. These are the lobby and the elevator. Apartment house lobbies are notoriously easy for anyone to enter. The main door lock usually has become so worn from continuous use that it can be opened with almost any key, inserted just far enough to turn the cylinder, or even with the edge of a dime. It's up to the tenants to see that such a lock is repaired or replaced.

Entry often is gained by pushing someone's bell, mumbling into the annunciator, and waiting for the click of the door release. In most buildings, even those with tight security at the front door, little attention is given to other means of entrance such as the basement garage. Most apartment garages are unattended, poorly lighted, and there are vast spaces from which hidden muggers can spring upon unwary arrivals. The garages also provide a means of access into the building through corridors or elevators reaching the garage level. One measure to deal with these problems is the installation of electrically operated overhead doors at the garage entrance, controlled by key or radio signal. This is effective only if the tenant driving into the garage closes the door immediately — if he waits until he has parked the car, someone can easily slip inside. Automatic, self-closing doors with photoelectric controls only partially overcome this deficiency. Attempts to solve the problem of elevator access by locking the elevator doors and closing off the stairways make it necessary for tenants after parking to walk around outside to the front lobby, and some object to this. Keys to the elevator doors may be the next step in this security measure.

Prowlers hanging around inside a darkened lobby are bent on robbery. Tenants can insist that the lobby be well lighted at all hours. That will reduce the possibility of intrusions and permit a quick reassuring glance by the tenant returning home to determine whether there is any unfamiliar and suspicious-looking person loitering there.

If the lighting is not at the usual level, take it as a warning. It would be best to beat a temporary retreat, find a telephone and call either the superintendent or someone of your family to meet you in the lobby.

A small whistle may be some defense. Easily held in the hand, it could be used quickly if an assault is threatened — a sharp blast on the whistle may well drive away an attacker, and likely will bring neighbors to their doors to investigate.

Before entering a self-service elevator, hold the door a moment while you look to see who may be there. Nearly always, any additional passenger will be a neighboring tenant whom you will recognize. Many elevators are fitted with a mirror to reveal any person attempting to hide out of sight in a corner. If there is a stranger, and his appearance makes you suspicious, just back out and let the elevator go. If you do enter, move immediately near the control panel and notice the location of the alarm button. At any untoward move by the other occupant, you should be ready to press — and keep pressing — this button to summon help. While assistance possibly would be slow in reaching you, the mugger may fear that sounding the alarm can trap him in the building, so will most likely decide to make his escape pronto.

If you find yourself in an elevator with a suspicious-looking character who gets off at the same floor you do, and follows down the corridor, don't open your own door, but rather ring the bell of another apartment. The intruder may well respond to this by fleeing down the stairs. Of course, it may turn out that he was just visiting another tenant, and thus had a right to be there. Your embarrassment at disturbing the neighbor may be covered up with some little excuse; for example that you were expecting a package delivery and wonder if it had been left there. You might tell the truth about what happened, and that's certainly a lot better, but it's a gambit that you can't repeat very often.

Certainly there is no easy answer to the dangers that lurk in unattended lobbies and elevators, particularly late at night. Temporizing with various devices may get you out of trouble on occasion, but the better part of valor is to make more reliable arrangements. A good watchdog is an excellent companion on walks, a dependable assurance against being taken by surprise in a darkened street. Less effective is a small pocket-size alarm that can be held in the hand, which emits a continuous siren sound that will very likely frighten off an attacker. And if the hazard of late hours makes you decide to come home earlier, that is just another of the adjustments required in this high-crime period.

Central Office "Hot Line." Direct burglar alarm connections to central office stations have long been used for protection of offices, stores and other commercial places. Recently, because of the increasing burglaries and the greater concern for security by citizens living in high crime areas, central office service has been extended to cover apartments as well. The Holmes Electric Protective Company now provides an apartment intrusion detection service which combines a monitored door alarm unit and 24-hour armed guard assistance. The equipment consists of a control box with an on-off switch operated by a special key. You leave the unit "on" when you are out of the apartment, or are alone or asleep. If an intruder enters, a signal light flashes on the monitored control board and uniformed guards are rushed to your apartment from a nearby guard station. In addition to the door detector, a remote control button can be installed on a bedside nighttable, or other location in your apartment which can be used to bring prompt assistance in the event of any danger, or other emergency such as sudden illness. The Holmes people say that the mere presence of their Holmes Seal on the apartment door has reduced the incidence of burglaries in the protected apartments. However, a series of forgetful door openings, without first disconnecting the alarm switch, will bring an embarrassing rush of armed guards to your apartment. If this happens more than just once or twice, you might find the response less prompt after that.

What is best to do when you awake during the night and have a suspicion that someone has broken into your apartment? The safest action usually is to play possum, pretend that you're asleep, and hope that the burglar will leave you alone after he has filled his pockets. But that depends also on certain things. If the intruder definitely is in another room, and you can move without being spotted, and if you have a lock on your bedroom door, then you might try sneaking quietly out of bed to reach and slam the door shut, then instantly throw the lock bolt. The situation then has changed, and you have at least a few minutes to attract attention. First try to use the telephone to call police, screaming all the while for help. If you think the burglar will make an effort to crash the door, roll the dresser and other furniture up against it — each piece adds that much to the barrier.

Banging on heating pipes, screaming out the window, rapping the floor with your shoe heels, all can cause a considerable amount of noise and disturbance at night, something no burglar is comfortable with even though he may have taken his own precautions by determining that no one else is in the building, that your telephone is inoperative, that there's no alarm system, etc.

While there's always a chance that the prowler may be a violent maniac, it is unlikely. The odds are that he has broken in for the purpose of theft, and is anxious to get away quickly. All the time that he is there you remain an easy victim for assault, and any panicky, poorly timed move can arouse a violent response. So your attitude must be pragmatic — taking the step that gives you the greatest chance to escape injury in the particular circumstances. If you live in a building where there are friendly neighbors on whom you can count for help, and you're sure there's someone nearby, then an effort to get their attention would be justified if it can be done with at least temporary safety. If you can use the telephone surreptitiously, dial your local emergency police number, usually 911, or dial operator and say "emergency," then give your telephone number, your address, then name or apartment number, and finally whatever information you can about the situation, all in that order.

Laundry Room Safety. An especially troublesome part of apartment living is the laundry room. Usually located in the building's basement, where there is little supervision over comings and goings through the service entrance, the isolated laundry room has been the scene of numerous robberies and personal assaults on women attending their washday chores, often entirely alone in a vast room. Many apartment managers now limit the laundry room schedule to just a few hours a day, at the same time maintaining a closer security watch including use of closed circuit television, and the result has been a considerable lessening of crime incidents. Recent building designs provide for smaller laundry rooms on individual floors, or alternate floors, and this is much preferred as it eliminates the basement trip and shortens the time that the tenant is away from her unguarded apartment.

Wherever possible, women using the laundry should arrange to go in pairs, but this is a difficult arrangement for busy housewives to make. Every effort should be made to avoid remaining alone in the room for longer than is necessary to load the machines.

8

Family Security Room

If you are awakened and have reason to believe that there's a burglar in the house, what do you do? Certainly it would be worse than foolhardy to sally forth to investigate and perhaps give battle with an intruder, single-handed in the dark, not knowing where he may be hiding, whether he is armed, how desperate he is, what he is after.

A practical answer to such emergencies is provided by a "security room" which has been set up as a second line of defense. Just having such a retreat available will add to your peace of mind, and in an emergency, may be life-saving.

Whenever endangered, if you can't get out of the house, quickly retreat with your family into the security room, putting a solid locked door between you and a possible attacker. In this way, you will avoid emotional panic and ill-considered actions, which sometimes are more responsible for tragic consequences than the original cause of an emergency. Specific planned procedures, to be followed when danger threatens, are an invaluable safety measure.

Selecting the Room. These security arrangements would not in any way affect the usual, everyday use of the room, nor need they require any basic alterations or extensive equipment. The first step is to select the room that would be most advantageous. Since a dangerous situation is more likely to occur at night, a bedroom is the best choice, especially in two-story homes. This room should be central to the other bedrooms, those occupied by children or elderly persons. The "master" bedroom usually is the most logical one, since it is the parents who would respond to and take command in a danger-threatening situation. Any of the other bedrooms that are too distant to the central plan can be fitted with their own protective locks and communication or escape equipment.

In apartments, things are more compact, with less distance between you and an intruder who jimmies the door or a living room window, and also less space in which you can maneuver. The function of a secure room still could be utilized as stated above, if the bedroom door can be locked swiftly and smoothly. A bathroom also may be suitable, particularly if it has an outside window that can't be reached from the fire escape, but you should not count on always being able to slip to safety from the bedroom without being intercepted if the intruder is in the next room.

Setting Up the Room. The essential detail of the security room is a dependable door lock that will function instantly, since locking the door might have to be done in a race with the burglar before he can get into the room. The old-style mortise lock, with which most interior doors are fitted, usually isn't adequate for this purpose; its "skeleton" type key works haphazardly at best, and tends to fall out of the keyhole, so it may be missing when needed. Securing the key with a chain or twine may hamper turning, and keeping the key on a separate hook will mean loss of time while fumbling it into the keyway. In any event, if such a lock is used, it should be kept lubricated and occasionally tested to make sure it is in smooth working condition.

An improvement, leaving the original lock intact, would be to replace the key with a fixed turn knob that comes complete with a new escutcheon plate. This knob is easily installed, and the cost is nominal, about $1.00. Such a unit is manufactured by the Safe Hardware Division, Emhart Corp., Item No. 1120 or No. K, 02581-3, available at most hardware stores. A separate mortise latch bolt, the type that is recessed into the door, is the most favorable solution. The Sears model 9P 57151, with key cylinder permitting the locked door to be opened from outside, or 9P 5989 without the cylinder, are suitable types. Standard exterior door locksets are even better for this purpose, requiring just the turn of an inside button to deadlock the door.

Reinforcing the Door. The condition of the door is important. The stronger the door, the more positive will be the safety precautions. If yours is a lightweight panel type, less than 2 inches thick, consider reinforcing it with an additional panel of ¼" or ½" plywood, laminated with rubber cement. This extra thickness stiffens the door, makes a great difference in the resistance to attack, and has an added value that you will appreciate in its sound insulating qualities, making for a quieter home. One drawback of the laminated paneling, however, is that the plywood edges are exposed. This has been overcome satisfactorily by the use of wood filler and paint, or by covering that narrow edge with metal channel moulding. Make sure, also, that the door frame is in solid condition, the latch strike plate securely attached to the jamb, and that there is no open space between door and frame that would permit inserting a screwdriver or prybar.

When the reinforcing panel is laminated to the outside face of the door it will be necessary to shift the position of the door stop mouldings. This step will not be required if the panel goes on the inside surface.

The room should have a window facing the street, possibly one that permits an easy exit in the event of emergency, by some means such as a rope ladder. Another important detail is to keep an extra front door key in the room so that you can toss it down to police officers or others coming to your aid, thus facilitating a quick search of the house.

Avoiding an Attacker. The strategy is to bring your family into the room as quickly and quietly as possible, and immediately slam and lock the door. It will also help to push

a heavy dresser or chest against the door; the burglar is unlikely to crash this unless he is after a specific item about
which he has some information, perhaps a particularly valuable piece of jewelry. But meanwhile, you have your own
options to forestall further invasion and attack—you can try
to get help, sound an alarm, or make your escape.

Use the telephone to call police or neighbors for help. If
the phone is not working, you then resort to whatever means
are available—which should have been decided upon beforehand in planning the security room—to arouse nearby neighbors, who would call in the alarm and come to your aid. A
whistle, or just shouting, can achieve the desired results.
Remember that raising a rumpus, while you have the chance,
may be enough to drive the intruder away.

An electronic bullhorn, which has a sound range of over
300 feet, is an excellent and handy device, obtainable from
Lafayette Radio for $8.95. The D type batteries operating
the horn should be replaced periodically to assure full volume
of sound. Another effective attention-getting method, which is
generally overlooked, is the use of firecrackers. The boom of
a cherry bomb at night could wake the neighborhood and
scare the daylight out of any nighttime sneak. Just a string of
tiny crackers also would do the trick even where houses are
situated quite far apart, since their machine-gun staccato
carries well at night and is not likely to be mistaken for other
sounds. The firecrackers can be set off harmlessly outside a
window if a suitable bracket is available to hold them in safe
position. It's just a detail, but a significant one—make sure
there's a book of matches handy in the room to set off the firecrackers.

Youngsters often connect up sending and receiving tickers
with low-voltage bell wire to adjacent houses of friends, for
Morse Code practice. Such a linkage, which usually has call
bells at either end, can be very useful also for adult communication when an emergency arises.

Panic Button. Another effective device is the panic button,
sounding a siren or automobile horn located somewhere outside the house, under the eaves. A horn obtained from a

second-hand or dismantled car can be easily hooked up to a transformer with a low-voltage bell wire. Also, any house alarm system with a central control box and separate bell or siren can be wired to include a manually operated switch, similar to a doorbell button. Pressing the panic button actuates the alarm even though the burglar has managed to disconnect or bypass the other sections of the alarm system, if the panic button is wired directly to the bell circuit. A similar panic button is included in central service systems such as those provided in certain cities by the Holmes Electric Protective Company and the American District Telegraph Company, over leased telephone lines. Several such buttons may be included in a home system, with one button located in the security room.

Distress Signals Helpful. With the vast increase of burglaries and personal assaults, many cooperative neighbors have developed ingenious methods to signal distress and a call for help. One instance of such neighborly communication that attracted much attention in the newspapers at the time, and had a happy ending, involved a manager of a suburban branch bank. Early one morning, three armed gangsters broke into his home as part of their plot to rob the bank. Leaving one of their number to hold the wife and children at the point of a gun as hostages, the other two bandits forced the bank manager to drive them to his bank, open the vault and hand over sacks of money. But meanwhile, the next-door neighbor, a New York City policeman, had noticed a pre-arranged signal in the family's breakfast room. The neighbor unhesitatingly phoned local police, who rushed to the home and subdued the gunman there. Quickly surmising the full situation, the policemen called their headquarters and patrol cars were rushed to the bank, arriving just in time to grab the two bandits.

Lighting Up. Lighted rooms and hallways are to your advantage, so throw on all the light switches you can while dashing into the security room. Of course, the burglar can turn off the lights, or smash the bulbs, but light may drive him off. It is still better if you have the modern low-voltage switching system with a master control in the security room. This permits turning on all lights, including the backyard and driveway floods, and also the basement lamps, from the central control switch. Lighting up the security room also will make you feel more confident, and the fully lighted house will help attract the kind of attention that any intruder tries to avoid.

Wait Patiently. Once in the locked room, how can you check whether there actually is a burglar in the house, or whether he is gone and the coast now is clear, so that you can safely leave the room? If you've managed to alert the police and they have made their usual search, you can feel confident that the danger is over. The police are experienced in sizing up the possibilities, know where to search for a possible prowler, and will not leave the premises until they are satisfied that all's safe. If there has not been any checkup, it would be better to patiently wait it out behind the locked door as there may not be any safe and sure way to check. Resist the inclination to open the door and peek, as the intruder may be waiting nearby just for that opportunity. When you do come out, don't feel annoyed with yourself if it all turned out to be a false alarm — it's more intelligent to act on suspicion than to ignore it until there is a dangerous confrontation, when it would be too late to protect yourself. Nor should one or two false alarms make you less responsive to any new possibility of danger; on the contrary, the one time you relax normal caution may be critical. The better attitude is to regard the initial responses not as jittery flights, but rather as trial runs, each one making you more capable of dealing with any real crisis should one occur.

REQUIREMENTS FOR A SECURITY ROOM

1. A solid door with a quick-functioning deadbolt lock on the inside.
2. Window or other means of emergency exit.
3. A means for climbing down safely, by easily-attached fire escape ladder or permanently anchored knotted nylon rope with hand grips.
4. A telephone, preferably a direct line, but an extension will do.

5. A secondary means of signaling, if the telephone does not work, such as hand-operated siren, loud whistle, firecrackers, bullhorn.
6. A fire extinguisher. Foam type, which is suitable for any electrical fires.
7. Ideally a master switch on low-voltage relays controlling all the house lights, or at least the outside floodlights, which would be turned on to help police nab the intruder.
8. A front door key in the room to let in police.

FIRE PROTECTION

If the fire warning signal is sounded, the recommended procedure is to immediately guide your children out of the house, and stay outside. If your exit is blocked by flames or smoke, then take the children into the security room, shutting all doors on the way, and leave by the window using your stored ladder equipment. If there is time, put in an emergency phone call to the firehouse, otherwise make the call from an outside alarm box or a neighbor's telephone. Don't delay your escape to gather up any clothing or valuables — it's dangerous.

The fire extinguisher is not intended to encourage you to fight the fire alone; however, some small fires may be caught at the very start, before they have had a chance to spread, in which case the use of an extinguisher may be warranted and may succeed in quenching the fire before much damage has been done. Even so, the fire should be reported to the fire department, which will take proper steps to make sure that the flames will not flare up again. Many a small blaze that seemingly had been extinguished turned into a major fire after smouldering undetected for hours.

Leaving by the Windows. Compact folding escape ladders have been developed for quick and safe emergency exit from a window. One all-metal type made with steel chain and aluminum rungs, has tubular brackets that are instantly and securely settled over any window sill. Tubular spacers steady the ladder and hold it away from the house wall, making for easier and faster descent, even for persons who would be nervous on any ladder. This ladder, rated to hold 1,000

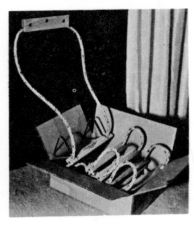

Emergency ladder should be kept secured at window and family should have drill in its use. It will hold several adults at once. It packs neatly in its own case.

pounds, enough for two or more persons to descend simultaneously, is available from Spartan Sales Co. or J. & P. Products Corp., priced $14.95 for two-story houses, and $21.95 for three-story houses, plus $2.00 mailing charges. The ladder rolls up into a small bundle for storage and easy access. In use, the support bracket is slipped over the sill, and the ladder simply allowed to drop down.

A true rope ladder, made of marine rope with oak rungs, is offered in the 14-foot length, enough for most two-story house windows, at $17.95 by Hotchkiss Products Co. Nylon strap and rope ladders, fitted with 1¼" aluminum rungs through which the side straps are threaded, and with attaching loops at the top, are available in any length on order from Rose Manufacturing Company. The nylon ladders are lighter, more compact when rolled up, and stronger than those made of hemp, and are not subject to deterioration when stored for long periods of time—all important details for emergency escape ladders.

THE SECURITY CLOSET

Always considering that an intruder may get into your house despite all precautions, you can still stay a jump ahead in

Security closet houses safe for valuables and also provides shelf space for guns and other dangerous articles that must be kept out of reach of children. Interior walls should have wire lathe and two layers of sheet rock.

protecting your valuables by providing a specially planned security closet. The security closet, an auxiliary part of the security room plan, is fitted with a built-in wall safe, filing cabinets, and specially arranged shelves and drawers to keep jewelry, reserve cash, stocks and bonds, important documents, and such valuables as fine silver, art objects, expensive furs. While some of these items belong in a bank safe deposit vault, the fact is that often several of them are brought to the house for one reason or another and are vulnerable to loss. The security closet also will be useful for storing guns and ammunition, fishing reels, and dangerous medicines or chemicals, all safely locked out of the reach of children, and doubly protected from loss by fire or theft.

While the closet could be set up anywhere, it is best located in your combination bedroom and security room, thus removing your most valuable possessions from areas where they would be quickly found. The back part of a walk-in closet would be highly suitable for this, if the space could be spared. Often, a new closet can be built recessed into a wall that backs against the roof eaves, thus providing additional storage space.

Specifications recommended by the Schlage Lock Co.,

Wall-mounted safe can be set into a closet from the rear or from a side wall, fastened through the metal flange at top.

The door of this safe is recessed so the combination lock knob is flush with the surface, permitting the safe to be completely concealed behind a picture or other means of surface covering.

Lid-type safe box can be set into a cabinet, or even into the flooring for greater concealment. Safes are rated for fire resistance, with Underwriters' Laboratories or Safe Manufacturers National Association rating label. Desirable for protection of safe contents is a rating of 1,700 degrees for one hour.

are for walls of ⅝ or ¾-inch gypsum board over metal lathe, a metal class B door frame, a solid-core class B door 1¾ inches thick, with 4½ x 4½ inch hinges having nonremovable pins, also an automatic door closer, and two Schlage locks, the A80P0 with ½-inch throw, and the B-460P, keyed alike. In planning the construction, do not overlook the fact that closets have four sides, and that often the side walls are weaker than the door. That is one reason for the metal lathe, which is most tenacious stuff to break through, in addition to having extra fire retarding properties. Double-thickness gypsum walls are recommended, one layer placed vertically, the other nailed and glued across, horizontally.

A small, fire-rated safe is built into the wall for jewelry, cash, deeds, securities, and other papers. Such safes are available from Mosler Safe Company, the Schlage Lock Company, Marvin-Hall, and others. A low-cost asbestos-lined safe is available from the Nor-Gee Corp. at $39.95 plus shipping charges. A Sentry Personal Safe, 17½ x 24 exterior dimensions, with 3-number combination lock, is available from the Accountants Supply House price $117.95 plus delivery charges.

If a safe is not enclosed in wall or floor, it is best to disguise it. This model is an attractive night table.

A round safe placed into the floor may escape attention, particularly if covered with carpeting. When encased in cast concrete, which can weigh 1,000 pounds and more, the safe will be virtually immovable and most difficult to attack in the floor situation. Such floor safes, as illustrated, are manufactured by American Security Products Co.

Hiding Places. A clever hiding place is often a better defensive measure than the most burglarproof lock and safe. If for any reason you will not have a security closet, there are many suitable hiding places around every house or apartment that can serve the same purpose. Your ingenuity can be put to work devising some secret place of your own, including such possibilities as false ceilings in clothes closets, slide out wall cabinets with space behind them, spaces between ceiling joists in the basement or attic, removable floor boards or tiles, with spaces between the floor joists.

But don't fall into the error of stuffing valuables into cubbyholes with just wrapping paper or envelopes. More than once this has led to tragic consequences when a plumber or electrician, doing some work in the house, came upon the bundle and discarded it because it looked unimportant. Best to place such items in special containers, such as metal bond boxes, which always have a small lock, available for $8.95 from mail order houses, such as Spartan Sales.

ARMS AND THE HOME

One consequence of the soaring number of home burglaries has been the purchase by many homeowners and apartment residents of various weapons, including firearms, for self-defense. Angered and at the same time frightened at the thought of a sneak thief breaking into the home, they obtain the weapons with a determination to resist, and of course defeat any intruder. But these arms purchases are made frequently without adequate regard to the many factors involved: whether the persons in the family who might have access to the guns, or might be called upon to use them in a crisis, have the personal capacity and experience to do so safely and effectively; whether the particular weapon is suitable to the

Round floor safe installed in concrete or into a wood floor offers excellent burglary protection and fire resistance.

Safe to be encased in a cast concrete block with welded steel casing, for installation on the floor. Concrete alone weighs over 1,000 pounds in 30-inch height, 22-inch sides. Safe may be bolted to the floor with special anchors.

intended purpose, and also whether having such weapons handy in the home might spark unpredictable and tragic accidents. The last item is as urgent as the others, since personality traits have a heavy influence; quickly aroused tempers sometimes prompt unintended resort to weapons on matters that would otherwise tend to cool down of their own accord. Inept handling of a gun when confronting a prowler may provoke a retaliation causing injury or death that might have been avoided. Your best hope is that the burglar is anxious to make his getaway as fast as possible without inflicting bodily harm.

A study by the National Commission on the Causes and Prevention of Violence reported that there are presently about 90 million firearms in private possession in the United States, and estimated that "about half of American homes have a firearm." Other estimates put the number of such weapons as high as 200 million, of which probably a fourth are hand guns which are easily concealable.

To arm or not to arm? That is indeed a difficult question. Experienced law enforcement authorities strongly advise against trying to attack or capture a burglar in your home, unless you are experienced in handling firearms and trained in apprehending criminals. Avoidance is clearly the best defense. The safest course is to concentrate on secure barriers and alarm systems, to keep intruders out, retreat into a locked room should a prowler manage to break in, and stay there until you're sure all is clear or police help has arrived.

You may confidently believe that you would be able to get the jump on a prowler unawares, putting him out of commission with a well-placed shot. But can you count on events working out so perfectly? It usually turns out the other way around, with the intruder having the advantage. Don't think the nervous burglar will stand around waiting to be cornered. You can't possibly remain on guard around the clock. The sounds of forced entry that may awaken you will be too late, and what about stealthy breaking in that you won't hear at all? In any event, the burglar may be also armed, more aggressive and desperate than you are, ready to spring from some unexpected spot or take a shot at you first if he's endangered.

The only favorable light on the subject is that the number of robberies in the home, which means burglaries involving physical attacks, is relatively small. In most cases, the burglar is just as anxious as his victim to avoid physical encounters. The thief wants to grab up what he can, and get out as quickly as he can.

But what if you are taken by surprise, while away from your bedroom investigating a suspicious sound, or some similar circumstance? In that case, you have no choice other than to cooperate with the demands made on you, handing over your cash and jewelry, or revealing where they are kept, quietly, without making a fuss.

Your best hope is that the robber is anxious to avoid inflicting bodily harm, which would land him in greater trouble if he is caught. But you can't depend on that. Many householders have been mercilessly beaten and pistol-whipped, sometimes because the robbers were not satisfied with the loot they obtained, or they felt there were additional valuables to be extracted. Most times, however, physical attack is a result of refusing to cooperate, or attempting resistance. In any event, the homeowner or resident would be unable to reach or use a weapon, and would be perhaps unsuccessful in any attempt to use one if the weapon was right at hand.

The record shows that while few criminal attacks in homes are prevented by the armed resistance of intended victims, use of personal weapons under agitated circumstances has resulted in innumerable incidents of accidental or mistaken shootings, with tragic effects. Many will recall the event some years ago in which a prominent and wealthy Long Island socialite, awakened by sounds in the house and fearful that a burglar had broken in, grabbed one of the family's hunting rifles for defense. Still dazed from sleep, and in a panic because of a recent rash of assaults in the neighborhood, she fired down a corridor as the sounds approached. The shot killed her husband, who had been up late, reading.

In the summer of 1970, a Washington, D.C. youth who was

very proud of the sports car he had been given as a present by his parents, heard sounds outside in the early morning and thought that someone was tampering with his precious car, which was parked in the driveway. The youth got his father's rifle and fired a shot through the window, which he later said was fired only as a "warning." The victim turned out to be a 14-year-old newsboy on his rounds delivering the morning papers.

The Weapon Quandary. Ownership of firearms is both a personal and community problem. For the individual, this many-sided question must be resolved generally by weighing the dangers that may be faced, the capability for controlled and effective defense if firearms are available, the responsibilities that are connected with possession of dangerous weapons including proper storage that makes the weapon instantly available when needed but always safe from unauthorized hands, especially from children.

Dr. Milton S. Eisenhower, chairman of the committee which made the study of violence, has stated in his report that no American should be allowed to own a handgun unless he can prove a need for it and meets stringent personal qualifications. He has endorsed proposals for uniform state laws making possession of a handgun illegal unless the owner is licensed by local police, and for confiscation of all guns not so licensed. This sensitive issue continues to plague the country, with little progress toward a satisfactory solution. The growth of crime, fostered in part by the availability of weapons, has prompted more law-abiding Americans to obtain arms for their own defense; thus the proliferation of weapons becomes a prime factor in both cause and effect of rampant crime.

Tear Gas and Mace. Frequent mentions of tear gas in news reports of riot control actions have prompted a widespread notion that the chemical makes a suitable personal defense weapon, that the gas incapacitates an attacker but is harmless to the user. But despite the extent of criminal assaults in the country, many communities have outlawed the use of tear gas by civilians, and made possession of such gas projectiles illegal.

These restrictions have a basis. Tear gas in the hands of an untrained person is not a dependable weapon, having a very limited range and unpredictable action; its use may result in effects that are contrary to those intended, often failing to halt a criminal while causing serious and permanent injury to the user or other innocent persons. What is worse is the tendency to use the weapon at any supposed provocation, often mistakenly.

Suppose you were to spray a container of the gas at a supposed attacker, and because of your inexperience or nervousness, fail to aim carefully so that the full effect of the shot is lost. Instead of incapacitating the attacker, this action may anger or provoke him to extreme physical action.

Pocket-size Gadget. For quite some time, mail order advertisements offered inexpensive tear gas projectors, small enough to be carried in a pocket or lady's handbag, and in handy pistol or fountain pen form. The wording and illustrations with these advertisements seemed to give assurance that firing off the contents of this projector would ward off any attack by disabling the criminal. Actually, the quantity of gas contained in the shell could not be depended upon to obtain the desired result.

A number of events led to a legal clampdown. More than one exceptionally fearful woman has let loose a spray of the gas in an elevator just because another occupant looked "suspicious." In the confined area of an elevator, the gas would affect the woman as much as the possible attacker. The tendency to hold the gas gun in her hand in tense and threatening situations made it too easy for thoughtless, impromptu action. It might be claimed that some assaults would have been prevented if the victim had such a weapon readily at hand, but it is possible also that many of these incidents might have been provoked by the mere move to "draw" such an undependable weapon.

Knives and Other Bladed Weapons. These meet the same objections as with tear gas, perhaps even more so. The use of a knife implies close-up, arm-to-arm encounter. Unless you can catch the intruder unawares, which is unlikely, and are very adept at using a knife or hatchet, you increase the jeopardy to yourself if you fail to fell him with a single stroke or blow. The weapon then may be seized and turned against you, or the assailant may himself be armed and will not hesitate to use a gun if he is cornered.

Your Legal Rights and Responsibilities. Defense of the home is your right in common law. You may wound or kill a burglar who breaks into your home, and not be subject to legal penalty. But there are many qualifications to these rights: Force may be used only to the extent needed to ward off an attack, or to permit escape from a definite danger of assault. A mere suspicion that a person is intending to attack is not sufficient to justify extreme measures. Suppose a neighbor coming home late at night and intoxicated goes to your home by mistake and tries to open a door with a key. Would you be justified in firing a shot through a window, wounding or killing him, under the impression that he was trying to break into your home? Suppose an apartment neighbor lost his key and was trying to get into his apartment down the fire escape, but mistook your window for his own and tried to force it open. Extreme actions, such as shooting, in those circumstances could be considered hasty and unwarranted, possibly involving you in criminal charges and perhaps also with a suit for civil damages.

Similarly, invasion of your grounds by children, even if they were to break into the basement on some lark or other, would not necessarily justify inflicting physical injury, since it might be held that you were not at any time in danger of an assault which was beyond your ability to repel by ordinary means.

The law is complicated and complex. You're not likely to start looking up law books, or call a lawyer, to find out what are your rights when danger threatens. No doubt you want to prevent physical injury to yourself and family, and protect your home. But you also need to avoid legal involvement for injuring or killing anyone, even a burglar, if you can help it. Then heed these basic cautions: make your house as safe as possible from invasion. Avoid physical confrontation with an intruder. Be cautious and restrained about resisting with any weapon.

9

Pros and Cons
of Burglar Alarms

The value of an alarm system is obvious. Just flipping on the master control switch gives you the comforting feeling that comes with knowing you have an electronic shield covering every accessible door and window. Housewives, particularly, appreciate the sense of security afforded by the alarm system; many women who had become increasingly fearful for themselves and their children because of the wave of residential crimes are able to relax for the first time in years when their homes are equipped with intrusion detector circuits. A fire alarm, in combination with the burglary detection, provides another vital protection, guarding the sleeping family from a stealthy menace.

But having an alarm system in the house can be a considerable nuisance, too. It's not all a one-sided benefit; the operation of the alarm makes its own demands, and it's either adjust to them or go without the security that the alarm system can provide.

Homeowners in the process of making up their minds about

135

installing an alarm system frequently ask these questions: How much does an alarm system really contribute to security? How does an alarm system fit into the everyday activities of the average family? What routine attention does the system require? Can a system function properly with children in the house? How extensive a system should be purchased? Is the expenditure worth while? Does the system get out of order frequently, and how difficult are repairs? Is a monthly service contract advisable? And lastly, what defects should be avoided in selecting an alarm?

The Electrical Watchman. There's no doubt that an alarm system can be a valuable adjunct to any home security plan. The mere existence of the alarm, indicated by an outside bell box, aluminum foil on the windows, even just a warning sticker on the door, may deter attempts at entry. If a door or window is forced, sounding of the alarm often is sufficient to scare the burglar away.

The alarm signal may attract the attention of cruising patrols, or neighbors may call police. If you arrive home and find the alarm ringing, it may save you from walking in on desperate burglars — best to call police who will check out the premises before you enter. When you are at home, the alarm signal alerts you to the presence of an intruder, so that your family can phone police and lock all bedroom doors or retreat to the security room.

Zoned Circuits. These provide continual coverage for certain parts of the house while the normally used doors and windows are released for family requirements; thus, only certain doors and windows need be kept under personal observation, while others remain under control of the detector devices even during the daytime. The latter protection is of increasing importance when the family is occupied with TV watching. Another example: guests may be entertained in the recreation room while the doors and other parts of the house are kept on alarm alert. Just throw the proper zone switches, and you know those entrances are safeguarded.

Oversights Corrected. When locking up at night, any door or window left unlatched will be at least covered by the alarm detector. In a related sense, the alarm system will show instantly whether any window or door has been overlooked and remains open. (The panel may be wired so that a "ready" lamp glows to indicate that the system is in readiness—with all doors and windows properly closed—for throwing the "on" switch. This will help avoid needless sounding of the alarm bell by switching on the circuit when a window is still open.)

Panic buttons are considered by most women to be a real blessing. Placed at the entrance door, the button can be used to sound the alarm should an intruder try to push his way into the house, or make any kind of attack. The panic buttons, similar to ordinary bell buttons, are usefully located at each entrance door, also in the kitchen, and particularly in the master bedroom, to summon assistance when needed. If the alarm system includes a phone dialer, this button is an effective way to call police.

When children are left with a sitter, the alarm offers additional effective protection, either automatically or by panic button should an emergency arise, if instructions are left with nearby neighbors or police patrols.

Plenty of Contras, Too. Despite the unceasing efforts and ingenuity that have been applied to develop effective and convenient burglary prevention devices, an alarm system is no simple panacea for the tremendous problem of home defense. Some families just can't adjust to the electronic controls. Half the time they forget about arming the system, and thus are entirely unprotected. The rest of the time, it seems, they forget that the alarm is "on" and open doors right and left, sounding a series of ear-splitting false alarms. Obviously, false alarms not only antagonize the neighbors, they cause all alarm signals to be ignored, thus invalidating the usefulness of the system.

Living with an alarm system requires a certain degree of awareness and participation by every member of the family. What if children arise earlier than their parents and are eager to rush out to play? If the master switch is in the parents' bedroom, it is simple to turn off the alarm at that time. If that is not convenient or desirable, then a time delay switch may be the answer, allowing the door to be opened for a fraction of a minute, but keeping the alarm otherwise on alert. The alarm will ring, however, if the children open the door again from outside.

Another possibility is to have a cutoff toggle switch, located on the frame of one door, which the children can be taught to use to disconnect that door's detector from the alarm circuit.

Shunt Switch at Door. A similar problem concerns the member of the family who returns home late at night, after the alarm has been set. A key switch at the door will allow entry without sounding a false alarm, but the system must be turned back on again from inside. The shunt switch controls only the contactor at that particular door, not the entire alarm system.

A different situation exists when a resident neglects to set the alarm on leaving the house "just for a little while." This gives the ever-watchful burglar the easy chance he's looking for. One such incident concerned a noted surgeon whose house had been burglarized only a few months previously. As a result, a very complete and expensive alarm was installed, including foiled windows and direct dialer to the police station. Not long afterward, the doctor and his wife went out for a social visit in the neighborhood, not setting the alarm because they expected to return shortly and it was early in the evening. When they did return, however, they found the house had been burglarized again, involving considerable loss of jewelry and much cash.

A family with all adult members, or just an elderly couple living alone, will be able to cope quite well with the requirements of the alarm system if determined to maintain maximum protection, though that may mean getting the switch key and turning the alarm "on" and "off" several times in the course of a single evening. Much of this effort can, of course, be minimized by various convenience features such as the time-delay exit switch, and by concealing the master control switch so that it can be operated without a key because of its secret location.

Malfunction Problems. One other difficulty occurs when the alarm system does not report "ready" state, that is, one of the circuit wires is broken or disconnected, or a detector is not functioning properly for one reason or another, most probably because a plunger has become stuck. If the cause cannot be located immediately, the alarm cannot be armed and the house is unprotected. A zoned system has the advantage that that one zone may remain out temporarily while the rest of the locations are covered. Tracing the defect usually is not difficult. Wire breaks are rare, unless some recent activity in the vicinity of the wires gives cause to suspect that condition as the possibility. In most cases, a terminal connection will be found loose, or a tarnished wire fails to make contact. In supervised closed circuit systems, such malfunctions occur more frequently and may be traced to the relay, the power source or a terminal connection. Very often the reason is vibration of doors or windows. During the 1971 hurricane Doria, police reported numerous sounding of alarms as a result of vibrations caused by high winds.

The consequence of such disruptions in the alarm circuitry, as with frequent soundings of false alarms, is a tendency to neglect setting the alarm on occasion.

Requires Proper Handling. Ultimately the alarm system falls into disuse as "troublesome" and "a nuisance." Actually, it is neither. But like any other device, it requires correct maintenance and handling to obtain its benefits.

Obviously the unit will do no good if it is inoperative, either through defect or failure to turn it on regularly, and therefore may be harmful if the family has come to rely on the alarm system and has let down its alertness. It may be a "nuisance" to climb upstairs to reach the master switch every time the alarm is to be turned on when the family leaves the house, or to turn it off when the arrival of guests is expected. Note that both these "hardships" can be eliminated—the alarm switch can be controlled from any of several locations by installation of a simple switching relay at the control box, and the alarm may be deactivated at any door by means of a bypass switch, preferably with a keyed lock, similar to the outside shunt switch.

The answer to whether an expenditure for the alarm is worthwhile is, then, that it depends on the extent to which you would value the protection that an alarm system provides, and your willingness to meet the requirements for effective operation.

The amount that should or need be spent for a system depends largely on two factors. One is the type of house you have, whether it is located in a high crime area or would attract the attention of thieves because it has or is supposed to have valuable articles or cash, and therefore should have a really extensive burglar detection installation. The other factor is whether you are willing to assemble and install the system yourself, thus reducing the cost substantially, by as much as $1,000 or even more.

Residential alarm systems, complete and installed, run from $750 to $1,800, with the average at approximately $1,300. Homes in vulnerable locations or with many entrances requiring extensive window foiling, installation of multiple mercury detectors in awning type windows and garage doors, and possibly vibration detectors for protection against smashing through walls, will possibly bring the system installation to $2,000 and even more. In contrast, the homeowner who installs his own system would pay only for the materials, which could cost approximately $250 for an open circuit system in an average-size home, $300 for a closed circuit system, and a maximum of $500 for a combination burglary-fire system including most of the important convenience features. Here is a representative list of materials with approximate range of prices and total costs, for an open circuit combination burglary-fire detection system in an average-size, 6-room home. A closed circuit system will involve a slightly higher cost for the panel box that includes the supervisory relays, but other components cost about the same whether for open or closed systems.

QUANTITY	COMPONENT	PRICE RANGE	TOTAL
6	All-purpose door and window contactors	$1.50 – $2	$10
10	Magnetic window detectors	3 to 4	35
1	Control panel	30 to 50	40
1	Outside gong in box	25 to 45	40
1	Fire horn (indoors)	10	10
1	Door shunt lock	10	10
500	feet of wire, 20 gauge	7	7
1	Control switch	1.50	1.50
3	Panic buttons	1.00	3
2	Tamper switches	2.50	5
2	Insulated pull traps	2.50	5
1	6-volt lantern battery or transformer	5.00	5
6	Heat sensors	3.50	21
2	Rate of rise indicators	8.00	16
		Total	$208.50

Smoke detectors range from $17 to $70. Incidental items probably will bring the total to about $250.00. Additional features will add to the above cost estimate. The most expensive is a phone dialer, programmed to call police or fire stations. Prices for this unit vary considerably; one supplier (Sunset Equipment Corp.) offers the complete two-track unit at $72. Other prices range from $250 up to $600 for various electronic units.

Installations requiring window foil will not add materially to the total cost, as the materials are not expensive. The metal foil costs about $4.50 for a 300-foot roll. The only other materials necessary will be a number of foil connectors, which are priced at about 25 cents each, some insulated brass strips, adhesive, and protector coating, all at nominal cost.

Is Outside Servicing Needed? A maintenance contract certainly is not necessary for the homeowner who does the installation himself, since he will be familiar with the materials used, the location of all the circuit wiring and the signaling equipment hookup. Perhaps in the early stages, some faults may show up, mostly poor terminal connections or wire splices, but these can be quickly cleared up and the system then should be trouble-free for long periods if care is taken not to damage the wiring or other components. A service contract which may cost $5 to $10 a month, may be useful when the system is a closed circuit type which is subject to more frequent and sometimes puzzling false alarms than is the open circuit system.

If the alarm system is purchased as a complete installed unit, then the contractor should supply a service warranty for a reasonable period, a half year to a year, without further charge. The homeowner would within that time determine by experience whether a monthly service contract would be necessary.

Avoid Shoddy Products. Quality standards have been established by various recognized organizations and government agencies for fire alarm equipment, making it possible to purchase needed materials with assurance that they will give dependable service. The Underwriters' Label (U.L.) on a fire alarm product is your guarantee that it conforms to the standards. Burglar alarm components, unfortunately, do not carry this simple quality mark. Purchase will depend on both the repute of the manufacturer and your own judgment of the product after careful inspection, although there are limitations on the latter capability. You cannot, for example, determine by visual observation the nature of the contactor metals, their coatings, temper, and other important factors. The safety of your purchase then will have to depend on both the reputation of the manufacturer and comparison with similar products — the overall design of the equipment, careful wiring, attention to details, all are indications that can be weighed in making a selection.

Useful Additions to Your Alarm System. With the basic control panel and circuit wiring, you can obtain important additional alarm services. These include: Flooding alarm, which will inform you quickly in the event of a burst water pipe (or the opening of an automatic fire sprinkler). Food freezer alarm, which signals when the freezer temperature rises above zero, or other preset degree, thus enabling you to take action to save your frozen foods from spoilage.

Power failure, by switchover to standby battery which sounds the warning that current is off and that appliances such as the water heater tank may be subject to freezeup damage.

These features can be included at very modest cost. For example, Allied Radio and Electronics offers a freezing alarm at $3.95 that signals when temperature falls below 32 degrees, a flood sensor for $5.95 that warns of water in the basement floor, and at $2.95 a power failure indicator that signals (on battery current) when the house power is cut off. These sensors are supplied with pin plugs, for connecting to the Heathkit Home Protection Panel, but can be used with other equipment.

It would be useful also to have a device that immediately warns of the condition of the telephone lines — whether the service is down because of central office trouble, a storm condition, or your telephone wires have been cut.

10

Alarm Systems and Components

You may start your alarm system quite small, covering just one or two doors and windows, expanding it anytime you wish to include other vulnerable entryways, adding fire alarm heat sensors, then putting in desired features to make the system all the more convenient and dependable. The latter include panic buttons, shunt locks, time-delay switches, and direct police dialers.

But first thing, you must make a definite decision on the basic type of system that you will install, whether it will be the "open circuit" or "closed circuit" type. That is because many of the components, particularly the detector switches, are not interchangeable. Your choice will be based on several considerations. You may want to have the most elementary — but still dependable — system, one that is easiest and least expensive to install, and simplest to operate without putting too many restraints upon the family activities. Or you may prefer a more complex and sophisticated "supervised" system that offers a greater assurance of security, but also costs much more, requires careful installation, has sensitive controls that occasionally are more troublesome because the instruments are so delicately balanced that false alarms can occur "at the drop of a hat."

Both systems are effective for perimeter security, and are compatible with all devices such as photoelectric cells and automatic dialers, except that window taping (foil) applies to the closed-circuit system. There are also many new electronic (solid-state) security systems, some with very limited capability but others with unique qualities that utilize sound waves, vibration, motion, radar, and other physical principles for detection technologies. Some of these systems are quite expensive and intended primarily for commercial and industrial installations. In the home, *perimeter* defense is the prime objective, sometimes backed up by secondary intrusion detectors such as the photoelectric cell, wired floor mats, and various traps.

The more simple and less expensive "open circuit" system is effective and satisfactory for most home installations. A "closed circuit" system is better suited for a family that consists solely of adults who can adapt to the requirements imposed by the sensitive controls. Following are brief descriptions of the two basic systems and some comparative data, to help you decide between them:

THE "OPEN CIRCUIT" SYSTEM

An "open circuit" system has detector switches of the "normally closed" type, that is, their contacts are separated when the button is depressed in normal position, as with the door or window closed. But opening the door releases the button and the contacts close, allowing current to flow through the wires to sound the alarm. All detectors in the circuit are wired in parallel: current flows only when any detector switch "makes" contact.

A disadvantage of this system is that it becomes inoperative if a wire is cut, a contact or terminal wire becomes loose, or some similar condition occurs. However, the circuit wiring can be concealed against tampering, and it normally cannot be reached from outside without setting off the alarm. But once the alarm goes off, it will continue to ring steadily even after the wires are cut because of a special device, called a Constant Ringing Drop. This device is connected directly to the alarm bell and bypasses the alarm circuit instantly when a detector has made contact. The constant ringing drop is fully described later.

The vulnerability of the open circuit system is minimized by a test button which sends current through the main circuit wiring and reveals any break in the wires; this test lights a small lamp or sounds a buzzer, but bypasses the alarm bell, which does not ring during the test. But this test does not show whether each individual detector is in working condition.

All it takes to set the open circuit alarm system is to throw on the control switch, which may be placed in any convenient room. False alarms would, of course, occur if a door or window is opened forgetfully by a member of the family while the alarm switch is still "on," but you can reduce this possibility by having cutoff toggles at frequently used doors. False alarms caused by malfunctions of the detectors or other components of the system are rare, and this type of system is relatively trouble-free.

THE CLOSED CIRCUIT SYSTEM

In the "closed circuit" system, low amperage current flows continuously from the battery to the supervising relay, through all the detector switches. These are the "normally open" type in which the contacts are closed when the buttons are in the normally depressed position, as when the doors and windows are shut against the switch buttons. The current from a $1\frac{1}{2}$-volt battery actuates an electromagnet in the relay, to hold a spring contactor apart.

If for any reason the current is interrupted, even momentarily, the sensitive relay opens and releases the contactor, setting off the alarm bell. And the alarm will keep going until the control switch is turned off and the drop relay reset.

Current to the relay can be interrupted when a door or window is forced, or a similar break-in occurs. But it also can happen when contact is broken for an instant by rattling of the windows by the wind or even by a low-flying airliner, a break in a screen wire, a short in the circuit, or erratic operation of the sensitive relay.

When such a condition occurs, the alarm cannot be reset until the fault has been located and corrected. Dividing the system into zones, however, helps to spot the location of the difficulty and also allows setting the alarm for the functioning sections during the disruption.

The closed circuit system requires more expensive and sophisticated equipment, the installation must be precisely wired, the system maintained in perfect pitch for dependable performance. The chief problem, however, is the tendency of this system to sound false alarms, which at the least are annoying, and may even be disastrous and counterproductive. It has been observed that in some homes, a consequence of these events has been to neglect setting the alarm except when the family leaves for extended trips.

"NORMALLY OPEN" AND "NORMALLY CLOSED" SWITCHES

The trade designations of "normally open," and "normally closed," alarm detectors can be confusing to all but experienced installers. An explanation of these terms is supplied here:

The "normal" condition of a switch is when it is in a state of rest, that is, unenergized or uncompressed. Thus, a plunger switch that is "normally open" (N.O.) will be one in which the contacts are open (separated) when the plunger extends from the case. When the plunger is pressed inward, as by a closed door or window, the contacts close and the circuit is completed. A closed circuit system, then, requires such *"normally open"* switches.

For the open circuit system, you would want a "normally closed" (N.C.) switch so that, when installed and the button is pressed by shutting the door, the contacts are open and no current flows.

Think of the switch as it is held in the hand. If it is listed as normally closed, then its contacts remain closed until either the plunger is compressed, or a magnet pulls the contacts apart. And the opposite applies to the normally open detector switch.

When ordering alarm components, it might be a good idea to mention each time the type of system you are installing. Some alarm detectors are marked N.O. or N.C., but in others the type is merely shown on the envelope or package. They are not interchangeable.

So the rule here is: select the opposite—you ask for N.O. if you have a closed circuit system, and N.C. for an open circuit system. But, like all rules, there are exceptions. One exception is with the fire alarm detectors, another is window foiling. These detectors are not altered by compression or magnets: the units simply go up just as they are. Thus fire detector switches are always open circuit, used in open circuit systems. If you have a closed circuit system, do not connect the fire detector circuit to the supervising electromagnetic relay, only to the C.R.D. All combination burglar-fire alarm panels are wired for this arrangement.

"EXOTIC" ALARM SYSTEMS

Security technology has made tremendous advances in recent years, producing devices ranging from the most simple and inexpensive door alarms to an extensive array of solid state circuitry equipment that seems right out of science fiction. Among the latter are ultrasonic sensors that function with highly sensitive microphones; infrared detectors that spot an intruder by the change in radiation in a specified area; proximity detectors, including microwave motion sensors; a radar-like invisible ray unit; photoelectric cells that hide in wall boxes and look just like ordinary electric receptacles;

even a laser beam application and a geophone system that employs sensors buried in the ground to detect the presence of intruders within a specified area. Closed-circuit TV in apartment lobbies and home entrances as a means of positively identifying callers, and other advanced equipment are harbingers of even more unique inventions that can be expected in the days and years to come from electronics.

Detector Devices Improved. For the homeowner, of greater importance right now are the steady gains that have been made in the development and improvement of practical, low-cost devices such as window and door contactors, magnetic and mercury switches, aluminum foil connector blocks, and similar materials that solve the problems of installing basic perimeter protection of the home. Equally important is the availability now at moderate cost of advanced, pre-assembled burglar alarm control panels correctly wired with the necessary instruments, thus assuring quality performance of a do-it-yourself installation.

Power Source. Current for most alarm systems is supplied by a 6-volt battery, or by a transformer (cost about $5) connected to the house power service. A lantern battery or hotshot battery has a shelf life of a year and more. Lantern batteries cost only a few dollars and should be changed every six months as a regular maintenance routine to assure dependable performance.

Many supervised closed circuit systems use a 1½-volt battery for the relay, and regular 120-volt house current for the alarm bell, with 6 or 12-volt standby battery.

A transformer connection eliminates concern about battery failure, but there is always the possibility of a power blackout, (even a momentary voltage drop can trigger an alarm relay) and also the danger that intruders can make the entire system inoperative merely by pulling a main switch or shorting the circuit. An ideal solution to this problem is a combination of transformer leads and standby battery, with instant automatic

Instrument panel is the heart of the alarm system, in which all circuit wires, power and alarm bell conductors are connected. Panels vary in makeup according to purpose, range of controls, and type of service. This one includes a milliampere meter on front cover to show condition of battery. Control by key is not limited to the panel box alone, but may be extended by a separate control switch to any convenient part of the house.

switchover to the battery in the event of power failure. Many prewired control instruments have this arrangement.

The ultimate power arrangement is an alarm system powered by storage battery with a battery charger on the house current keeping the battery always at full charge. Thus the alarm system can function at any time independently of the home electrical current, and there is assurance that the battery will always be in operative condition.

Transformers should be wired directly into the main circuit panel, rather than plugged into a receptacle, since the prongs may become loosened or pulled out, breaking the current flow. In no event should transformer lead wires be plugged into a receptacle that is controlled from a wall switch, unless you deliberately wish to use the wall switch to turn the alarm system on and off.

BUYING THE MATERIALS

Savings in cost are often possible when all material for the planned installation is purchased at one time, as unit charges for single items usually are considerably higher than carton prices. Most suppliers require minimum billing for each order, and the shipping cost may be the same for larger packages as for small items. Suppliers usually will pay the freight for shipments totaling certain minimum amounts. It will be worth while to write to the supply houses for alarm component catalogues.

Quantities of the materials needed depend, of course, on the scope and extent of the system you are installing. The particular types of the various components also will depend on the individual situations in your home, the kinds of doors and windows, the degree of vulnerability of the entryways, and many other factors. A whole "library" of devices has been developed over the years to cope with every situation, and you will most likely find a solution to every problem among the contactors and other switches described in this section. If you run into difficulty with a unique problem, you possibly can get the necessary answer by writing to one of the components supply firms listed in the appendix, describing the condition.

Basic to the whole alarm system are the detectors — those devices that stand guard over all entryways into the house, and trigger the alarm signals when intrusion is threatened — or accomplished. They are illustrated and described here so you can select the ones that you will need.

Magnetic Detectors. The most effective, versatile and dependable detector is the magnetic contactor or switch. The device comes in two parts, each completely encased in dustproof, weatherproof plastic. One part is the complete switch; it has the two wiring terminals and the contact points, which are enclosed. The other part is the magnet, which is attached to the moving part of the installation, the door or sash. The parts are installed closely together, two small screws in each.

Perhaps the most effective and dependable type of detector is the magnetic switch, which is highly versatile for installation arrangements. Part with magnet is attached to the sash, door, or other moving part; the switch part with wire terminals is attached to fixed position such as the window frame or door jamb.

When the sash is raised or the door opened, the magnet moves away and thus shifts the internal switch points. One type of this switch comes with normally open contacts, which become closed when the magnet is engaged. It is used for the closed circuit system. A similar type is for open circuit system. (They cannot be intermingled in any one circuit.) The individual detector units, available in either white or gray color, are 2½ inches long, one half inch wide, and 9/16-inch high. These switches cost approximately $3.50 each.

The magnetic detectors can be used on steel casement windows with an insulating block under the magnet. The detector parts in most instances must be attached to the movable steel sash by drilling and tapping the metal for small holes. The switch part is fastened to the stationary part of the window, or on the frame, if sufficiently close to the magnet.

A metal screen used over steel casement windows may inhibit use of a magnetic detector. One alternative is application of lead foil to the glass, linked to the frame by means of a foil connector block, another is installation of a plunger-type switch such as the tamper switch, the plunger extending through the metal frame so it is moved by the sash.

An important detail to watch when purchasing these magnetic detectors is that there are two distinct types. The original type, called "reed" magnetics, has the moving parts sealed in a glass vial within the plastic case. Cracking of this glass, which can occur for any number of reasons — such as dropping the switch before it is installed or damage in shipment — would break the hermetic seal and expose the electrical contacts to corrosion, resulting in unreliable performance. There is no way to determine visually the internal condition of the detector.

Another shortcoming of the reed type is that the steel contact springs provide only slight tension and thus the contacts may fail to make adequate connection when released by the magnet, or the contactors may "freeze" into one position. The reed type detector is best suited for locations where there is frequent action or "exercise" of the contactors, such as at a door to monitor the comings and goings of family members.

An improved type, developed by the Alarm Device Manufacturing Company (Ademco), omits the vulnerable glass container. The contactor springs have more than three times the tension of the reed type, thus providing more positive pressure on the contactors. Additionally, the silver-coated contactors operate with a wiping action that assures a smooth, non-pitted surface even though the switch may remain nonoperative for very long periods at a time. The Ademco magnetic switches, which have become the standard for this type, are somewhat higher priced than the reed type, and the distinction should be noted when making a purchase.

Tamper Switch. Another device that can be used for casement swing-out windows is a tamper switch which has a long rust-proof plunger. The switch body is encased in nylon plastic. Attached to the window sill close to the sash so that

Tamper switch with long plunger is useful at casement window where magnetic detector would have to be insulated from the metal window frame. Closeup shows three holes for mounting screws and terminals at back for wires.

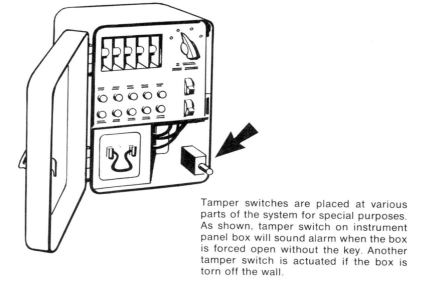

Tamper switches are placed at various parts of the system for special purposes. As shown, tamper switch on instrument panel box will sound alarm when the box is forced open without the key. Another tamper switch is actuated if the box is torn off the wall.

the plunger is depressed when the window is closed, the circuit is made or broken (according to whether it is a closed or open circuit switch) when the sash is moved outward. The tamper switch is an Ademco product, No. 16 and 19, for ⅝-inch plunger.

All-Purpose Contact. This is a shallow beveled button that is used primarily on doors but also is suitable for double-hung windows. The button is installed on the hinge side of a door frame, recessed into a ¾-inch hole, and attached to the frame surface with screws into the thin brass flange. When the door is closed, the button is depressed. These all-purpose detectors are available from a number of alarm supply houses and are quite easy to install, by snaking the wires down inside the door frame or by removing the door trim to fit the wires inside and down through a hole drilled into the floor.

Button Contactor. Used mostly for double hung windows, the button is set into the frame under the bottom sash so that

The all-purpose detector is a shallow beveled button with very thin flange for mounting on door jamb (at the hinge side) or on windows. A ¾-inch hole is drilled to recess the body of the switch, while two screws through the brass flange holds the switch in place.

Button contactor is sometimes installed into the sill of windows so that the plunger is depressed when sash is lowered, thus readying the alarm. This switch also is used for auto theft alarms.

the contact is depressed when the sash is lowered. Installation requires drilling a ⅝-inch hole for recessing the button, then a ¼-inch hole for dropping the wire down to the floor below. Mechanical switches are not as dependable as the fully enclosed magnetic detectors, as moisture and dust at the push button entry can affect its function, causing the button to stick in closed position.

These buttons may be used also for manual emergency or "panic" switches. The buttons are available from Protecto Alarm Sales and from any electrical supply firm for about $1.25 each.

Toggle Switch. This is a standard inexpensive item, available at most electrical and electronics supply houses, including Lafayette Radio and Allied Electronics. The toggle is used for a master control switch, cutoff switch, and similar applications.

Entrance Shunt Lock. Entrance to the home while the alarm is "on" is accomplished with the shunt lock, which is installed on or alongside any entrance door and turns off the alarm. The lock is threaded into a steel sleeve which contains a lock

Toggle is used as control switch in simple alarm systems, or for shunting a segment of the circuit, such as at a frequently used door, when the rest of the system will remain armed.

cylinder. Keys are removable in both On and Off positions. Only the detector on that door is controlled by the shunt lock. Various types are available. Prices range from $7 to $15.

Constant Ringing Drop. This essential component of every alarm system, open or closed circuit, is described more fully in Chapter 12. Every alarm instrument panel includes some

Key shunt lock on door frame permits switching off segment of alarm system to permit entrance or exit without setting off the alarm bell. An indicator light shows when the alarm is on.

Shunt lock in barrel housing with removable washers at both ends. Lock has shielded terminals to which the circuit wires are soldered. Installation may be on the door frame or on the door itself.

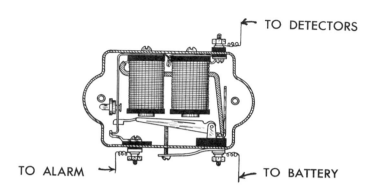

TO DETECTORS

TO ALARM →

← TO BATTERY

Constant ringing drop is an essential element of the alarm system. This relay is always included in any pre-wired instrument panel. Diagram shows details of the C.R.D., which is used in both open and closed circuit alarm circuits. This relay should be included in the home-assembled alarm system. Current applied to coil through the two opposite terminals energizes the magnets, attracting the armature. The lower contact spring drops and touches the lower contact. Current from the transformer or battery now goes directly through this contact to the alarm bell, bypassing the detector circuit. Bell continues to ring until the C.R.D. spring is reset.

type of C.R.D. When the detector device energizes the circuit, even for just a moment, the relay drops a spring contactor to close a direct circuit from the power source to the alarm bell, which will ring continuously until the drop relay is reset by hand. Thus, if a burglar were to shut a door instantly on slipping inside, the alarm would not be stopped, but rather would continue unabated until reset. Therefore, the control box containing the constant ringing drop should be located in a hidden place or locked closet, or in a tamperproof metal box. The constant ringing drop can be purchased at any electrical supply store, and at some of the larger hardware stores that have electrical departments. Prices range from $8.00 to $15.00.

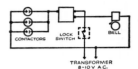

In open circuit system, detector switches are wired in parallel. One wire lead goes on single terminal side of the constant ringing drop, other wire is spliced to one transformer lead and continues to the alarm bell. Second transformer wire goes to constant ringing drop terminal opposite the one from the detector, while another wire from the third terminal connects to the bell. The control switch is located on the transformer wire leading to the C.R.D.

Detectors of a closed circuit system are wired in series. One wire, spliced to a transformer lead, connects to the constant ringing drop; the other wire goes to the relay. The second transformer lead connects to both the relay and the alarm bell. Single conductors are run from the relay to the constant ringing drop, and another from the C.R.D. to the bell. The control switch cuts into the circuit before and beyond the detector locations.

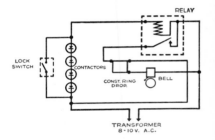

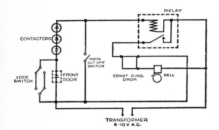

Closed circuit system with zoned detectors allows a door or window detector to be made inoperative for usual family activities, while the rest of the system stays on alert. Lock switch in line bypasses a section.

Leaf Springs. These come in a number of types and sizes, adaptable to applications where other devices will not do satisfactory service. These springs are useful on awning type windows and similar difficult locations. One type of leaf spring designed for home installation is only 9/16-inch wide and 2 inches long, has a tiny spring contact.

Plate Contacts. Each 3/4-inch wide, these plates are mounted face-to-face on doors and windows to carry circuit current

Spring contact, one of the original alarm detector devices, is still useful for certain locations, particularly at screens, storm window inserts, and where a gap exists at door closures.

Leaf switch operates by pressure on the extended arm above the switch housing. It is useful in situations where other types of detectors won't fit.

through double contact points. Used extensively for telephone work, this standard contactor can solve some installation difficulties.

Door Trip Switch. This is used at the top of a door. It has an offset spring-loaded lever that is moved each time the door is opened. Easily installed and dependable in performance, its chief defect is the appearance, as it projects above the top of the door.

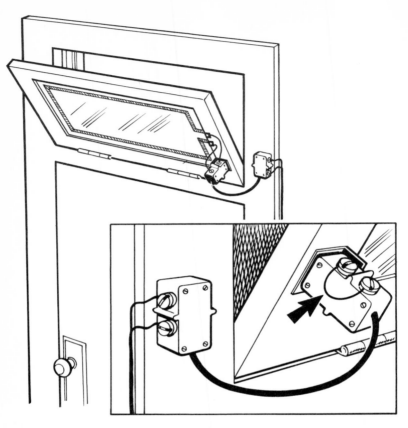

Overdoor transom has flexible cord from the foil takeoff block, so that the transom can be operated normally. Mercury switch on transom or awning window functions when tilted beyond specified limit.

Adjustable Mercury Contacts. Equipped with a heavy-strand copper cord connecting the two blocks of this device, the detector can be installed to allow extensive movement. Used for doors and awning type windows, pivoted casements, transoms, and similar installations. The internal mercury switch is sealed in unbreakable metallic block. Available with cords of 12 and 18 inches.

Vibration Detector. This alarm is used for safes, filing cabinets, cash drawers in commercial establishments, also suitable for protection of art collections and other valuables in the home. The vibration detector can supplement other burglary protection installations by providing a response to hammering, sawing, and other efforts to break through a wall. The miniaturized detector switch is adjustable to the required sensitivity to avoid false alarms initiated by extraneous or intermittent vibrations common to that location, such as the passage of planes overhead.

Wire cord provides flexible connection at door from soldered foil block on pane to fixed wire terminal.

Vibration detector is a compact unit that can be mounted to a cabinet or wall, will respond to unusual or excessive vibration that results from efforts to break in or tamper with a safe or filing cabinet lock. An adjustment screw at the back permits control of switch sensitivity to avoid false alarms from normal activities.

Floor Mats. Pressure-sensitive mats, wired to make or break contact when stepped upon, are useful adjuncts to the alarm system. The mats are so thin that they can be placed under regular carpeting or rugs at any locations that need to be protected — under windows, at doorways, and on stair treads, for example. Mats may have single or double connecting wires for either open or closed circuit system installations.

The mats are sold by all firms supplying burglar alarm components. Lafayette Radio & Electronics offers several types that can be ordered by mail, including one 17 x 23 inches, catalogue no. 33F66010, price $12, a stair tread 6 by 23 inches, catalogue no. 33F66028, at $4.90, and a runner five feet long that can be cut to any required length, catalogue no. 33F66077, price $18.40.

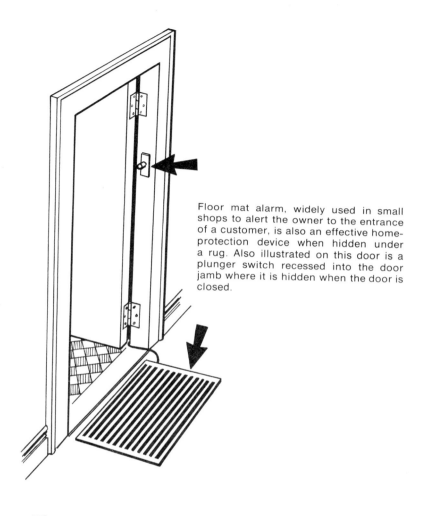

Floor mat alarm, widely used in small shops to alert the owner to the entrance of a customer, is also an effective home-protection device when hidden under a rug. Also illustrated on this door is a plunger switch recessed into the door jamb where it is hidden when the door is closed.

Ultrasonic Alarms. These devices, only recently available, may well be the forerunners of the ultimate in burglar detection systems. They can detect and scare off the prowler even before an attempt at entry is made, which is the best kind of protection. The flexibility and ease of installation also are favorable attributes. However, there may be some shortcomings in that alarm soundings would occur whenever a dog or

Typical ultrasonic alarm unit is contained in small case that looks like a table top radio, plugs into a nearby lamp receptacle. A person entering the supervised area will intercept the rays. Immediately the plugged in lamp will ight up. After a stipulated time delay, perhaps 10 to 15 seconds, the alarm horn sounds, and continues even after the person has moved out of the range. The time-delay feature gives the homeowner a chance to shut off the unit before the alarm goes off if he has to enter the area for some reason.

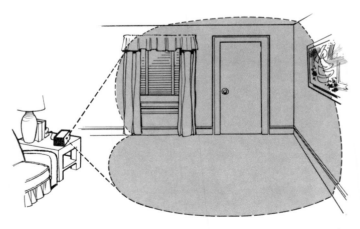

Shaded area shows how ultrasonic waves spread out to form a cone within which any movement sets off alarm.

cat wanders into the detection beam. There is a 1-minute delay after setting the unit "on" allowing plenty of time to leave the house with the alarm on guard, and a slight time delay after the alarm device is triggered.

An ultrasonic alarm is manufactured by 3M Company, called Intruder alarm, another by the Artolier Lighting and

Back of ultrasonic alarm shows the on-off toggle switch, range control knob, and terminals for remote alarm bell which can be linked with low-voltage wire. The remote bell can have its own constant ringing drop and operate on battery power, so it can't be shut off from the ultrasonic box.

Sound Division of Emerson Electric Co., under the name Alert-Alarm. The unit, plugged into any 120-volt house receptacle, has terminals for connecting a standby battery and for remote alarm bell. Range can be adjusted from a few inches to a maximum of 30 feet. A further description is provided in Chapter 7 for apartment resident service.

Traps at Work. There's always a practical solution if you run into a seemingly hopeless complication, when none of the regular detectors can be applied to a particular situation. Problems like these may arise with jalousie windows, with aluminum sliding windows where it is difficult to attach half of the magnetic switch to the metal sash, and some bathroom windows surrounded by ceramic tiles which cannot be disturbed.

Trap cord switch is adaptable to many situations where other types of detectors may not be suitable. Examples are across basement windows, at doorways, stairs, any entranceway, and large glass areas such as a greenhouse. The trap switches may be included in either open or closed alarm circuits, but a special type is required for each.

In such locations, one practical solution is a trap wire with switch clip. One wire can protect an average size window, and two may be used, if needed, to leave no more than 12-inch clearance on larger openings. If entry is attempted, the trap wire is pulled out of its spring-loaded retainer and rings the alarm. Installation is nearly always quite easy. The trap wire may be placed either vertically or horizontally across the window, in front of a door so that opening the door pulls the trap wire, or even tied to a Venetian blind or window shade. In fact, an entire home system can be installed using only the wire traps at all locations.

Overhead garage door is linked into the alarm circuit with a flexible wire cord in two sections, joined by interlocking plugs (arrow). When door is raised above a certain height, plugs fall apart and break the connection to sound the alarm.

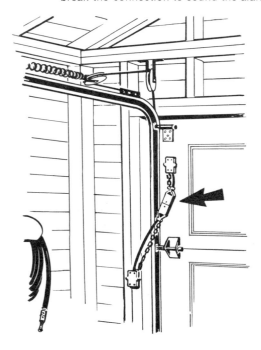

Installing the Trap. The usual installation is with the trap wire suspended from a bracket at the top of the window frame, at center or best position for maximum functional value. Directly below, attach the trap itself. Attach the wire to the clip, adjusting its length so that the wire is just taut, and tighten the wire retainer screw on the clip. To set the trap, slip the clip into the spring-loaded grippers.

For open circuit systems, an insulated trap is used with a fiber wire clip. The circuit wires are connected to both sides of the trap. When the clip is pulled out, the trap contactors spring together to close the circuit and ring the alarm.

For closed circuit systems, the single circuit wire is connected to the top bracket. The circuit continuity is obtained by the trap wire which is a conductor, and the same circuit wire is continued from a terminal at the gripper end. Pulling on the trap wire will break the circuit and ring the closed circuit alarm. The traps are available from all burglar alarm supply houses.

Self-contained trap unit that is plugged into a nearby electrical outlet has both burglar and fire detectors with built-in alarm siren.

Trap (A) consists of opposing steel balls in contact under spring pressure. An inserted trap clip (B) is held in position until withdrawn by opening of door or someone tripping over a floor cord. A fiber clip keeps contactors apart for open circuit system, a metal clip on a continuous wire maintains current flow for closed circuit systems. The wire is held taut by a spring hanger (C).

"Panic Button." This permits sounding the alarm bell manually, independent of your detector circuits, in the event of any home emergency. Such a button installed near the front door can be used if any intruder attempts to push his way inside, or enters the house by some ruse. Another such button placed in the master bedroom or the security room can be sounded in case of fire, entrapment, or serious illness, when there is no other way to summon assistance.

The entire system becomes more functional with a shunt switch, which is a key-operated cutoff switch located outside at the entrance door. Using a special key, a family member returning home late can enter without setting off the alarm or having to awaken the household. An adjunct to this shunt switch is a small red light that glows when the alarm system is "on" as a reminder not to just open the door—and also as a warning to thieves. The shunt lock permits setting the alarm when going out, as the key disconnects the system while the door is open, then the switch is returned to "on" position from outside when the family leaves.

Automatic Telephone Dialer. This electronic supplement to the home alarm system automatically calls any pre-programmed telephone number or series of numbers (to police or fire department, local patrol service or other emergency service) and relays a recorded message to bring help, stating

Automatic telephone dialer automatically places a call to police if burglar alarm goes off or to fire department if fire alarm sounds.

the location and nature of the alarm. The message, on a cassette magazine, is played by a tape recorder which is coupled to the telephone line.

The self-dialer, which has been approved by the telephone companies and does not interfere with normal use of the telephone, may be programmed to deliver 10 and even more messages in sequence to different numbers, or make repeated calls to the same number until the unit has been shut off or the cassette tape completed.

The telephone idea works best when linked by private leased line direct to the police station, as dialing may run into the difficulties that sometimes are encountered by phone users.

The dependability of the dialing device is seriously compromised if the telephone wires leading into the home are exposed and vulnerable, easily clipped with pliers. Any dialer system should be backed up by encasing these wires in metal conduit.

Telephone dialer instruments are manufactured or sold by the following: Diebold, Inc. of Canton, Ohio 44702. Scientific Security Systems. Dialalarm, Inc. Alarm Devices Mfg. Co. and Bourne Electronics.

Photoelectric Sensor. A light projector and a receiver are the basic components of the photoelectric system. Interruption of the light beam sets off the alarm. The system is designed for indoor operation. Protection may be extended to a larger triangular area by using reflective mirrors. The receiver has a photo cell relay which operates from a 1½-volt No. 6 battery, which has a service life of one year. The trans-

Photoelectric cell is a useful alarm instrument that can be set up quite easily to guard an entrance, a stairway, or any specific area. The unit consists of two separate parts, the light beam, and the receiver which sounds the alarm when the beam is disturbed. Units should be solidly mounted so that the light beam will not stray off the focussed receiver. Guard keeps unit from being knocked off alignment.

mitter uses regular house current, or a 24-volt line from a transformer, with standby battery in the event of power failure.

Photoelectric Relay Kits with effective distance range up to 50 feet in darkened areas or nighttime use, and 15 feet for daylight operation are available from Allied Radio, Chicago, at $19.95. A complete photoelectric set is available at $19.95

Compact photoelectric detector looks like typical electric receptacle, fits into standard wall box. Shown here is the receiver, which is set across the room from the light source, contained in a similar fixture.

from Lafayette Radio & Electronics, catalogue no. 99F0821. Sensitivity can be adjusted to the precise beam power necessary for the distance between the units. The Ademco photoelectric system features a built-in meter which indicates the correctness of beam setting.

A shortcoming of this device is that any shifting of the light source or receiver can trigger an alarm. Overcoming this problem is a photocell system that is mounted rigidly in wall boxes like electric receptacles, providing trouble-free performance and more esthetic appearance.

Other suppliers are: Honeywell, Design Controls, Inc. Alarmtronics Engineering, Inc.

Wireless Remote Control Monitor. A radio transmitter which is capable of switching on a control relay up to 300 feet away, without wire connections, monitors a group of intrusion and fire detectors within a certain area. The control box has a radio receiver, and will respond to the signal from any number of the remote transmitters. The chief value of this system is that it eliminates much of the wiring that would otherwise be necessary to connect up an entire circuit to a central control. The battery-operated transmitter, which is only the size of a cigaret package, can be secreted anywhere in a room nearest to a group of window and door detectors. It also has a distress button which can be used in an emergency. A remote control unit, consisting of one transmitter and one receiver, is available from Lee Electric Co. at $165. A similar unit is available from Space-Guard, Inc.

Fire Alarm Devices. A fire protection alarm is, of course, an essential installation for every home. Fixed-temperature thermostat sensors, smoke detectors, and warning horns can be fitted into your burglar alarm system, on the same or separate wire circuits. It is important to note that the "normally open" type of heat sensors are used with both the open and closed circuit systems. Full details on fire alarm installations are given in Chapter 16.

Transformers. Step-down transformers convert house current to the lower voltages required by an alarm system. They may be connected at the fuse box or plugged into any house receptacle. Transformers are available from 6 to 18 volts at 5 watts, or 8 to 16 volts up to 25 watts. Large alarm bells or horns may require the higher wattage current, and this can be checked by reading the metal plate on the device. In some advanced alarm systems, standby batteries are used with automatic switchover in the event that power fails or has been cut off by an intruder.

SOME TRANSFORMERS

| Sears Roebuck | 34P 1405 16 v. 15 watt | $3.19 |
| | 34P 1406 10 v. 10 watt | $2.49 |

Ademco	#86	8 v. 5 watts	NA
	#87	6-18 v. 5 watts	NA
	#88	8-16-24 v. 25 watts	NA
Lee Electric Co.	#323	16 v. 15 watts (surface mount)	$3.45
	#327	8-16-24 v. 25 watts	$5.90

11

Installing the Burglar Alarm

First step in installing the alarm system is to put up the panel box. This can be placed in any out-of-the-way location — inside a closet, in the basement, even up in the attic if there will be access to it when necessary for periodic testing of the circuit wires and power source, also to reset the relay after an alarm has been sounded. The panel box — of metal or wood — may contain the control switch if that is convenient; otherwise the master switch can be located at any point in the house — preferably in a bedroom. Pre-assembled panel boxes are fully described later in this chapter.

All the necessary wires for the circuits start at the panel box, or the battery which usually is located in the box. It is possible to bring individual wires from each detector switch to the panel, but that will result in an unnecessary and cumbersome duplication of wires. Good technique, as practiced by commercial alarm installers, is to keep all or most detector switches on a single run of wire whenever possible. Careful planning will make this possible, carrying the wire from one switch to the next. This is for the basic circuit; items like the master switch, bell ringing lines, and battery or transformer connections are carried separately.

Layout Board Helpful. The homeowner who is not thoroughly experienced in tracing electrical diagrams will find it helpful to set up a circuit layout board before starting the work. In this way, the wiring can be visualized and easily followed while doing the actual installation. The little time spent setting up the board will be well justified by helping to avoid confusion and errors.

A piece of plywood about 3 x 3 feet in size would be most suitable for this purpose, providing plenty of space for arranging the various detectors, control switches, manual buttons, shunt lock, and fire sensors that will be used, all wired into a miniature circuit as shown in the photograph.

The setup includes a buzzer connected to a transformer or off-board 6-volt dry-cell battery so that each segment of the circuit can be checked out. The circuit wires are held in place on the board with staples, and each detector switch is marked for the place where it will be located, such as "kitchen door" or "basement door," etc., so that the board layout can be more readily followed later. Incidentally, this step in the project is instructive and also may be entertaining for those without previous interest in electrical wiring or electronics.

Types of Wires. Use No. 18, No. 20, or No. 22 stranded wire. The 20-gauge is preferred as it is heavy enough for easy con-

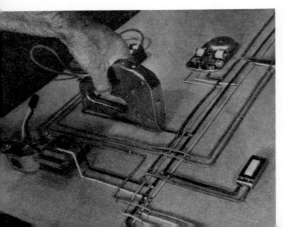

Board layout of alarm wiring and components may be helpful to homeowner who lacks experience in electrical wiring and finds it difficult to follow schematic drawings. The layout permits rearranging the circuit in accordance with the home's specific construction, making actual installation easier and error-free.

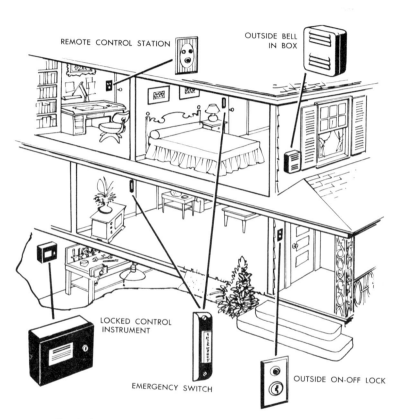

REMOTE CONTROL STATION

OUTSIDE BELL IN BOX

LOCKED CONTROL INSTRUMENT

EMERGENCY SWITCH

OUTSIDE ON-OFF LOCK

Controls for burglar-alarm system may be installed at strategic points about the house, as shown.

When many electrical connections are to be made a wire stripper becomes an essential tool. This is a typical style used by professional electricians.

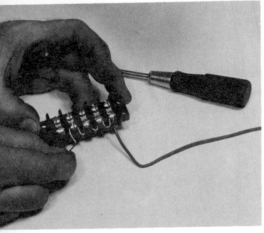

Terminal block is very useful adjunct to alarm components, as it permits connecting segments of the circuit into the block so that just one set of wires need to be carried into the instrument panel.

nections to the terminals, flexible and light enough to go through small wall openings. Wires should be of uniform size throughout the system, except for those leading from control panel direct to the alarm bell, which can be No. 16 or No. 18, typical lamp cord wire.

Twisted two-conductor wire, with insulation in different colors or markings — or the wire leads tinned for polarity identification — will be easiest to handle for terminal connections, although the linked two-conductor wire is better for long runs through walls and closets. At some parts of the installation, 4 or 5-conductor cable will be useful, reducing the need for snaking two or more separate wires through the same area of the house. Clear plastic insulation is quite attractive and suitable for locations where the wire must be in exposed positions, such as over base shoe moldings and around door frames, which sometimes can't be avoided.

A complete installation in an average size home will consume quite a lot of wire, often 300 feet or more. Wire is purchased most economically in full reels of 500 feet, and you may be able to buy the full reel at the same total cost as would be charged for just half that amount when it has to be measured out at retail prices.

Instrument Panels. Pre-assembled panels greatly simplify alarm installation, improve the efficiency and dependability of the system. The panels assure proper engineering and hookup of the critical elements of the system, and also provide for additional facilities that make the alarm more functional, including such features as circuit and power test buzzers,

Panel box may be placed inside a closet or out-of-the way place. Attach solidly with the heavy screws provided. Shown here the Ademco closed circuit, combination fire and burglar alarm panel.

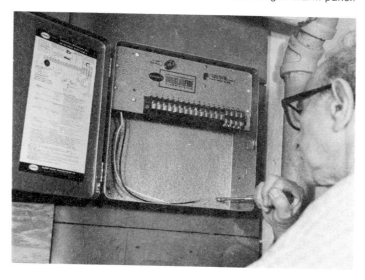

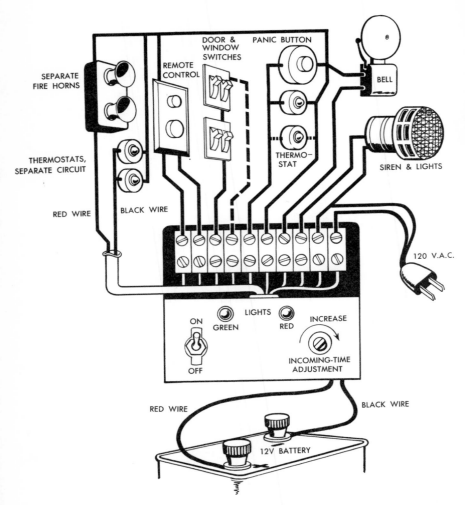

Typical panel detailing sheet, showing positions of the various connections to terminals. This part of the operation is greatly simplified in the diagram which shows wires merely attached to indicated terminals. This is an open circuit system panel.

battery meters, "on-off" signal lights, zone arrangement of detector circuits so that part of the system remains always on alert while other sections are turned off during daytime use. Wiring for these many details becomes exceedingly complex for persons who lack extensive electronics experience. But with the pre-wired panels, the connections involve merely attaching the necessary leads to the marked terminals.

The instrument panels are not very expensive, and may be purchased either separately or in kits which also contain the necessary components to complete the system. Most panels are for either burglar or fire alarms, but some are fitted for a combination to cover both hazards and include a large bell for the burglary circuit, and a warning horn or siren for the fire circuit. Batteries are not included, since they have limited shelf life.

Zoned Systems. Zone control, including pilot lights to pinpoint the source of an alarm signal, is extremely useful on the closed circuit system when it is necessary to isolate a condition causing a false alarm.

Separate zone circuits make it possible to keep sections of the system on alert when other parts of the house are released during the daytime for family use. An advanced instrument has a meter that shows when all circuits are closed. Thus, the bell is automatically by-passed for the moment when the alarm is turned on at night, and the bell does not ring if a window or door has been left open, thus avoiding an unnecessary false ringing of the alarm and permitting the correction without disturbing the family.

Many of the control boxes are equipped with tamper switches which guard the box against sabotage or vandalism. There is even a combination control panel and signal box, containing the alarm bell and its batteries, which can be mounted outside the house in areas of mild climate.

As an example, a panel and kit that would be satisfactory for most home installations is the Ademco No. 5160A. The panel, for open circuit systems, provides for separate burglar and fire alarm circuits, connected to different sets of terminals.

The burglar circuit will sound an 8-inch underdome alarm bell, mounted outside the house in a louvered metal box; the fire circuit will set off a distinctive fire warning horn.

Battery in Panel Box. The panel box has space for a 6-volt lantern battery, which is connected to two designated terminal lugs, while two other terminals are reserved for connecting the leads from a remote master switch. Emergency panic buttons and door key switches are cut directly into the circuit at the detector locations and not wired into the panel box, but the kit contains instructions for these installations also. There is no limit on the number of detectors and other devices such as photoelectric relays or floor-mat switches that can be connected into this or any of the other preassembled panel boxes.

For a "supervised" combination burglar and fire alarm system for homes, there are closed circuit control instruments which operate from 115 volt house current, with automatic switchover to standby 6-volt battery in the event of power failure.

Wiring the Circuits. Open circuit and closed circuit devices are wired differently. In the open circuit, the normally closed switches are wired in parallel, that is, both wires are connected to the switch, one wire on each terminal. The same wires then continue on and are connected in the same way to the next switch or device. There is no need to follow polarity for this type of circuit. The way many commercial installers make the connections is to split the linked wires with a knife or razor, then skin perhaps half an inch off the insulation of each wire and loop it around the terminal screw. Thus the

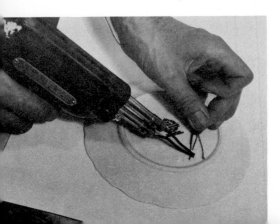

Stranded wires often cause difficulties because the loose strands break off or tend to make contact with other wires. Soldering is quick and very easy. Do the soldering on a fireproof surface, such as a piece of asbestos or an ironware plate. No flux will be needed on the freshly stripped wire.

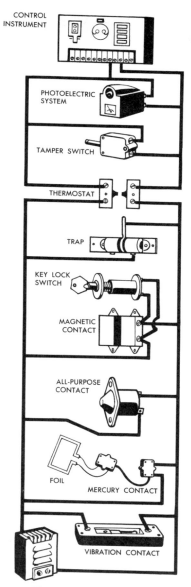

CONTROL
INSTRUMENT

PHOTOELECTRIC
SYSTEM

TAMPER SWITCH

THERMOSTAT

TRAP

KEY LOCK
SWITCH

MAGNETIC
CONTACT

ALL-PURPOSE
CONTACT

FOIL

MERCURY CONTACT

VIBRATION CONTACT

POWER PACK

In the open circuit the elements of the system such as detectors and switches are wired in parallel, that is, both wires are connected to each element, one wire on each terminal.

wire has not been cut and there's less chance of an interruption occurring. The uncut wire is continued to the next location and connected in the same way, but the wire simply ends at the final switch — no return is needed as the two-conductor wire serves as the complete circuit whenever the contacts close on any one of the switches, forming a bridge across the line. Of course, you don't have to use continuous wires, just so long as the ends of segments are in firm contact. Also, there's an advantage in bringing both circuit wires back to the panel box, thus providing a means for testing the circuit outside the alarm bell linkage.

Cutting in Branch Lines. There are circumstances where it is difficult to carry the wire both ways to a detector — in and out again. In that case, simply tap off a branch run of the wire to the detector, and continue on with the rest; just be sure that the splice of the thin wires is secure, and the only way to be sure is to solder it, then insulate with plastic tape. Also, always tighten all terminal screws firmly. It's a good idea to scrape the stripped wire so it is metal-bright before making terminal connections.

There are some devices that are wired differently. Step-on signal mats are wired for open circuit systems with one set of connecting leads. Both conductors are connected to the 2-wire lead at one side of the mat, while the circuit wiring is continued from connections to the leads at the other side. The mat responds to tread pressure, closing the contacts to sound the alarm. In this instance, it is wiser, though not necessary, to maintain polarity. Watch the code colors on the mat leads and follow through on the opposite end.

Closed Circuit Wiring. Normally open detectors are wired in series. A pair of twisted wires starts at the instrument panel (but do not connect to the terminals until wiring is completed) and is brought around the circuit to all the detectors in the system. But only one of the wires is cut, skinned, and attached to both terminals of the various detectors. The other wire remains intact. Polarity need not be observed except when called for on double circuit devices that require four connec-

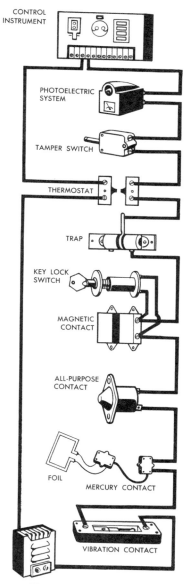

CONTROL
INSTRUMENT

PHOTOELECTRIC
SYSTEM

TAMPER SWITCH

THERMOSTAT

TRAP

KEY LOCK
SWITCH

MAGNETIC
CONTACT

ALL-PURPOSE
CONTACT

FOIL

MERCURY CONTACT

VIBRATION CONTACT

POWER PACK

In the closed circuit the switches are wired in series, that is, one of the two wires is connected to both terminals of each switch and the other wire remains continuous throughout the circuit.

187

tions, such as on floor mats, double circuit foil on glass, or window screens. In any of these, connect both conductors to the two wire leads of the device, and connect the new end of the circuit wire to two exit leads of the device. But this time polarity must be carefully maintained throughout—the positive conductor going in is reconnected through the same coded lead to the positive conductor of the continuing circuit wire. Then continue the circuit wiring to the next detector, resuming the original procedure of one wire connected to both terminals in series, the other wire left intact. At the end of the run, and with all doors and windows closed, bring the circuit wire to the panel box and connect the free ends of the wires to the instruments, which in most setups will be one side to the relay, one to the master switch or as marked on the panel diagram.

Window Foil Methods. Aluminum foil used in closed circuit systems serves as the conductor when cemented to window glass. Any breaking or cutting of the glass will break the circuit and set off the alarm. This system is used for large picture windows which cannot be otherwise protected, on door panes, and wherever detector switches are not practical.

Metal foil comes in rolls, ⅜-inch wide, and about two-thousandths of an inch thick. A roll contains approximately 300 feet of foil, usually enough to do a number of windows, and costs about $4. The foil is applied with special glue or varnish, but there also is available self-sticking foil at slightly higher price, said to go on as easy as masking tape.

Crossover strips are used in the foil applications. These are insulated brass, designed to simplify connections between

An example of window fully protected by metal foil applied to glass surfaces. Foiling provides maximum protection, as any efforts to bypass the alarm system by breaking a window will be detected. Note the insulated brass connectors leading from one glass pane to another.

mullions and crossover bars of steel casements. Also, there are special self-adhesive connector blocks for linking the foil with the circuit wiring.

The application starts by marking the glass with tailor chalk, or china marking pencil on the reverse side, using a length of wood or straightedge in the desired pattern and making sure that the lines are level. Clean the glass thoroughly. Attach the adhesive-backed foil connector blocks on the sides of the window frame at the foil positions.

Apply adhesive with a brush only as wide as the foil (3/8-inch), covering only small sections at a time at first, until you become accustomed to handling the delicate foil. Allow the glue to become very tacky, then start to apply the foil, unreeling it from a nail or dowel held through the center of the roll, or make a plywood reel. Apply in a continuous strip starting at one edge of the glass but leaving a 2-inch-long tab at the end that can be folded over to double the thickness where it will enter the connector block.

At each corner, start the fold in the opposite direction, then bend it back over the fold to form a neatly squared corner, and continue on without a break. Smooth down the foil as you go with the back of a matchbook in a squeegee motion. Repair any breaks in the foil with small patches, but without applying new varnish—rather make tiny holes at either end with a pin and cover the patch with varnish.

Foil must not touch the metal frame or crossbars of the windows, as they would short out the circuit. Apply the insulated brass crossover strips over tape fastened with varnish to the metal, for continuing the foiling. The sketch shows the various ways that this is done, and also the soldered connections to the circuit wiring, using the foil take-off blocks. Screw-type take-off blocks eliminate need for soldering.

Excess varnish can be removed from the window with benzene. The foil can be protected against wear and damage by applying a coat of Foil Protector varnish.

The Constant Ringing Relay. An essential component of the burglar alarm system is the Edwards constant ringing drop. This device keeps the bell ringing once contact is made,

Applying Foil

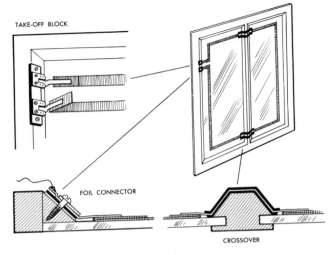

TAKE-OFF BLOCK

FOIL CONNECTOR

CROSSOVER

Window foiling procedure is outlined in the sketch. Metal foil is available in self-adhesive type, or applied over brushed on adhesive. In one method, a tinned brass shim, attached to plastic takeoff block with screw, is soldered into doubled over foil at point of connection into the alarm system. Brass shim with insulated section, is used also at points where continuity is carried across window mullions and other obstructions over which foil cannot be applied. These can be seen in the photos of the foiled window.

no matter how quickly the door or window is closed; without it the bell would just ring for a second. But the C.R.D. will keep it going and the bell cannot be shut off until the control switch disconnects the current. If you have a pre-assembled control panel, it already contains the constant ringing drop and you won't have to make any connections to it.

The device consists of an electromagnet and contact spring, and has three wire terminals. When current actuates the magnet, the spring contact connects the direct line from the battery to the signal bell, bypassing the detector circuit. Only by resetting a button in the C.R.D. when the current is off, is the contactor restored to its normal position.

Wiring the constant ringing drop may seem a bit tricky but is really simple, as seen in the accompanying diagram. Of the two-conductor wire from the battery or transformer, one

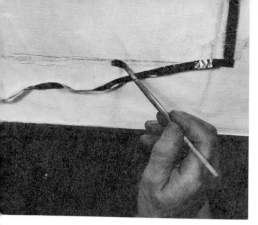

Application of the foil is quite simple, although a bit of practice is necessary to achieve a neat result. Application of foil is shown here against opaque background to emphasize details. First step is to mark out the foil pattern by snapping a chalked line over the glass. Apply adhesive on only one strip at a time.

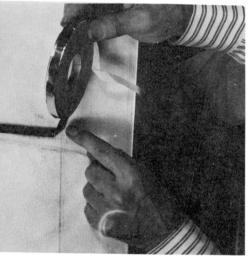

Foil is unwound from roll, smoothed down with a strip of cardboard so there is firm contact and a neat appearing surface. At end of strip, bend foil diagonally in opposite direction to that which will be followed to form the mitered corner.

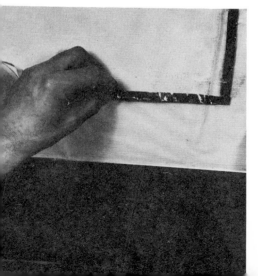

Fold the foil back over the first fold, to continue the turn, making sure that the corner is at true right angle. After applying adhesive, continue the line. Handle gently to avoid breaks.

191

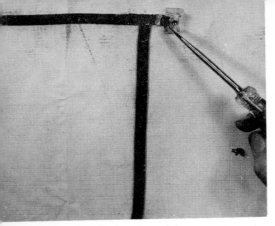

Foil is connected into the alarm circuit wiring by means of a takeoff block. This has an adhesive backing, so is simply attached to the glass at the end of the foil. The foil end is doubled over and attached to the takeoff block by tightening a screw.

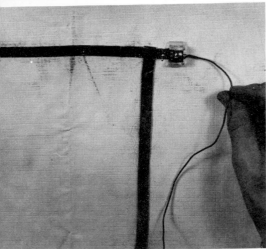

Circuit wire is attached to the takeoff block by connecting to the terminal screw. Where foil must pass over mullions, an insulated brass shim is used to make the jump.

wire goes through the detector circuit and is connected to the single terminal on one side of the C.R.D. The *same wire lead* is also spliced into a line to the alarm bell. The second wire from the battery goes directly to the C.R.D., which in turn has a connection from its third terminal to the alarm bell. When the relay contact drops, the current flows not through the detector circuit, but directly from battery to the bell.

Cutoff Switch at Doors. Convenience in "living with" the alarm system is assisted by the installation of a cutoff switch at the inside of any door that may be used frequently, such as one leading to the garage. Thus, it is possible to leave temporarily without shutting off the entire alarm system, by throwing the "off" toggle at the door, which cuts out the one location but leaves the rest of the system on alert. But don't forget to set the toggle back to "on" when returning. This becomes a routine procedure after a while. A time-delay switch overcomes the problem of the "forgotten" door.

The manual cutoff switch at the door is installed directly into a loop of the circuit wires to the detector at that switch location. For the open circuit system, the circuit wires go first to the cutoff switch, then to the detector, and on to the rest of the circuit. But remember that if the switch is "off" all the other detectors from there on in the circuit would also be off. To avoid this, a bypass is provided in the circuit by making the wire loop for that particular cutoff switch and door detector, spliced into the continuing circuit wire. Thus, even if the switch is left "off," the rest of the system has continuity and will function.

Shunt Switches. The door lock switch allows the alarm to remain on when leaving the home. The key-controlled switch does not turn off the entire system, but only controls the detector at that door, setting the contactor as required (open or shut according to the system used) when using the door. Therefore, the shunt lock is simply wired into the circuit at that particular detector, and the key either makes or breaks contact according to the position of the key slot. In practical wiring terms, the lock will cut off current from that door detector when the door is open, but will serve to maintain flow of current in the closed circuit system by the wire passing through the lock. It is necessary to set the lock before opening the door from the inside when the alarm is on, but any member of the family arriving home can disconnect that particular door detector, leaving the rest of the system functioning.

12

How to Conceal Wiring

The most difficult part of an installation is concealing the circuit wires that connect the intrusion detectors and fire sensors with the instrument panel and alarm bell. Of course you wouldn't want these unsightly wires exposed in the room around windows and along the walls, so every possible effort is made to snake them completely out of sight.

Fortunately, there are practical ways to do this, neatly and efficiently, in almost every kind of situation, and without causing damage to the home decorations. This is possible because the alarm system wires carry only 6 volts of current, the same as the door bell. The plastic-coated wires may be safely strung inside hollow walls, run through closets, stapled behind door trim and under moldings, and threaded through tiny holes in baseboards and floor boards from one floor to another. They do not need the protective boxes, conduits, or insulating devices that are required for the regular house electrical wiring.

No Code Restrictions. The only stipulations in the National Electrical Code regarding this work in homes are that these low-voltage wires not be run closer than 2 inches to the house light and power lines, nor be placed inside the same conduits. Where the alarm wires cross conduit or BX cable, they should not be allowed to fall across such cables, but rather separated from them by stapling to nearby beams or by placing pieces of wood blocks between the lines.

The techniques described in this chapter cover most situations that you are likely to encounter in the average home, and undoubtedly will stimulate your own ingenuity in solving any individual problems that arise. Installation is easiest in one-story homes where unobstructed access to the exterior walls and other framing is available from both the basement and the attic. Thus wires from the control box can be brought in every direction to the door and window contactors, while fire alarm heat sensors can most readily be installed in the ceilings with the wires carried along in the attic and snaked down inside an exterior wall or run through a closet, while the alarm bell connections are made by running them through the attic to an outside wall box. Another possibility—bring wires up at a doorway where they will be covered by the wood threshhold.

Two-story homes usually present some challenges, particularly when the basement room is paneled and some of the detectors must be installed on the second floor, as at a door leading to a deck over the garage, or a vulnerable window, also for alarm control switch and panic button in the bedroom. But even here, patient maneuvering nearly always can manage to work the wires through for full concealment. Only in rare situations are extreme measures required, such as cutting an opening into a finished plaster wall to get around some obstacle, and even here the work can be planned for minimal damage.

You can derive a great deal of satisfaction by taking advantage of every possibility for snaking concealed wires from place to place.

Special Tools Needed. In addition to the standard tool assortment (hammer, chisels, electric drill, screwdrivers, instantaneous soldering gun, awl, pliers, and a wire stripper adjusted to the gauge of the wire that you are using) you will need a few special tools. An essential item is a long ¼-inch bit: this may be an elongated bit, 12 or 18 inches in length, or an extension rod holding a ¼ or ⅜-inch bit.

Steel "Snake" Essential. The most important tool for concealed wire installation is a wire fish tape, or "snake," about 20 or 25 feet long. The tape is of spring steel, with flat surfaces ⅛ to ¼ of an inch wide, with sufficient tensile strength so it won't sag or bend too readily. Ordinary steel wire is not suitable for this purpose as it tends to curl around any obstruction. For two-story home installations, it will be helpful also to have an even longer tape available for reaching through the veneer or framed wall from attic direct to the basement.

Another useful item will be a length of straight and very stiff wire which you can make from a wire clothes hanger, the kind supplied by all cleaning stores. Just clip off the twisted hanger loop and bend the ends so that you have one straight piece about 18 inches long. This wire can be used when fishing through a partition wall from one room to another, or through the floor boards into the basement. The stiff wire will be easier to use and will locate drilled openings more readily than will the flexible tape.

Another device for snaking wires is a beaded metal chain, similar to the kind that is familiar as a key chain. Electricians use the chain in 12-foot lengths for locating openings inside walls. An end of the chain is dropped through the hole drilled into a window frame or door jamb, and the chain is jiggled up and down until it manages to slip through some open space,

Steel snake in a case is handier for the longer lengths.

Low-voltage wires are tacked down with special staple gun that shoots staples with rounded tops allowing free tug on the wire. Avoid driving the staples directly into the conductors or wire insulation; center the wire in the stapler guide before pressing the trigger bar.

perhaps the gap between floor boards and joists, or at heating and soil pipes. When the chain has dropped through so that the end can be reached, a thin steel wire is tied to the other end and pulled through. The circuit wire then can be drawn through the wall.

A useful tool is a *wire stapling gun.* One made by Arrow shoots rounded staples so the low-voltage wires are not damaged, yet are held in place. The stapling gun will help keep the wires neatly tacked along joists and inside closets, so they will be less likely to become snagged and the circuit thus damaged and made inoperable.

Each phase of the alarm installation warrants some study to determine the best method for concealing the wires to each of the detector stations. As there are considerable differences in home layouts and details, every home will present its particular variables.

Attic Flooring. One important precaution is to watch your step in attics that do not have flooring over the joists — take along a couple of plywood panels, 24 x 48 inches, or several long boards, for footing and kneeling support. Even in such "easy" situations as the attic, it is best to lay out a wiring plan on paper, detailing how you will approach each of the detector stations and showing the dimensions and location of each room.

For "problem" situations, such as rooms over crawl spaces and where attic areas are inaccessible, study the other possibilities. Keep in mind that every wiring difficulty has been encountered before and a practical solution has been worked out. The methods and techniques described here can be applied or modified to fit every condition.

Fishing with a Wire Snake. Using the fish tape to find openings inside blind walls for pulling the wires is more a matter of patience than skill, although some electrical workers become wizards at maneuvering the wire snake through seemingly impossible places. Patient effort is usually rewarding, as nearly always there's some tiny space in the framing or gap in the floor boards for the wire tape to get through — all it takes is about ¼-inch clearance. The tape steel is straightened, then bowed slightly near the end so that it will curve gently into the wall cavity.

Handling the "Snake." The tape is pushed along until it becomes blocked, then is maneuvered by twisting, turning and jiggling in an attempt to find a place for the steel wire to slip through. This should always be done slowly; otherwise the tape will be withdrawn even if it has found an opening, and the efforts wasted. This latter is a common error of the novice who is uncertain and impatient, and thus misses what would be a successful effort if he had been more persistent and watched the tape closely.

When necessary to bring the wires across from one side of a room to the other, the easiest method is to bring the wires back to either the basement or attic, tack them along the joists, and come back through floor or ceiling at the location of the door or window frames. But sometimes this method is not feasible, and going across the ceiling is the only way.

First determine the direction of the ceiling joists, as the fish tape will have to be sent in that way. A notch is cut into the plaster at the corner of the ceiling (remove the crown molding if there is one) so that the snake can be inserted, and another hole made at the opposite end to snare and pull out the tape.

Wire tape is used to draw wire down from attic through closet, or, as here, inside veneer wall. Bend end of tape into tight loop to take the wire.

Repair patches are usually not troublesome, as ceilings are nearly always painted plain white, and the plaster patches can be touched up easily to match. It won't be easy to work the snake across the ceiling joists, but it can—and has—been done.

Pulling the Wires. There are specific ways to draw the wire through after the snake has passed the barriers. The circuits should start at the control box, and the wire brought from there around to the detector stations. There's no law about this, and it sometimes is easier to start the wire at a detector and run it along to the box, but the former is standard professional practice and should be followed when possible, as it is the most expeditious method and helps avoid errors.

The springy metal tape is usually pushed through from the hole that has been drilled for the detector or heat sensor. One end of the tape is bent into a tight loop to be used for pulling the wire through the area that had been fished. The other end is left in original flat condition. But when the tape is sent on its search errand, the plain (unbent) end is inserted into the hole as it will be more able to pass through small open spaces. When the snake has gone through, and can be reached at the other side, say at the basement, you may be able to pull the wire back when it is folded over the steel tape loop.

Another way to do this is to tape a length of the wire to the straight end of the snake. Use masking tape, as that is thinnest and quite strong. Place about a foot of the wire alongside the snake, make the first winding of the tape only around the snake, very tightly, then wind down to cover also the wire and continue until about a 3-inch length of the wire is firmly secured. You've had a tough time getting the snake through

Wire is pulled down at end of snake. In tight quarters where a loop in the steel snake might not pass, simply tape a 6-inch length of the wire to the snake very tightly.

and you don't want to lose it now by letting the wire pull loose. Start pulling back very gently on the snake, not tugging it when it balks but rather trying to ease the wire through by several sharp jerks. If the wire won't pass, pull the snake all the way back again and check the masking tape. If damaged, retape and start over. There is every reason to expect that the thin wire will pass through with the snake if the tape is smoothly wound. Sometimes a coating of grease on the end helps.

A variation of this technique is possible, in which the snake is sent upward from the basement through holes that can be seen for guiding it. At the drilled detector hole, a short stiff wire (made from a piece of wire clothes hanger formed into a hook) is used to snag the wire and bring it out through the hole.

Wiring Door Detectors. The primary detector at a door is the contact button recessed into the hinge side of the jamb, so placed that the button is pressed in when the door is closed. This installation requires a 3/4-inch hole into which the housing is recessed flush, and the wires are attached to the shell at the back. This button can be located at any position along the jamb, but between 6 and 12 inches above the floor may be best if the wires are to come up from below. It is also possible to bring the wires down into the door frame from the attic, but this will be more difficult because of both the additional distance and the framing headers across the top of the doorway.

Nearly always there is a clear space of several inches back of the door jamb, between the jamb and the framing studs. When the hole is drilled, it may be possible to work the snake down right through to the basement, as there is no floor plate beyond the side studs. If this fails, because there are spacer blocks or the flooring boards extend under the doorway, then drill a 1/4-inch hole from the basement at that position. The drilling may be done more precisely from above using a long flexible bit. It will then be quite simple to pass the wire snake down through to the basement, bring up the wire, connect it to the button terminals, and attach the button to the jamb.

Wiring Door Contactor

Wing bits are used for drilling larger holes, such as this one of ¾ inch diameter for recessing a door contactor.

Wire is drawn up with snake at wall opening made accessible by removal of door trim. A little plaster of Paris will patch the replacement like new.

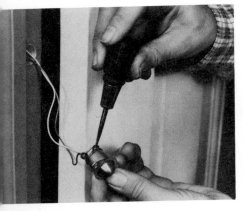

Wire is brought up through the drilled hole with snake and attached to the contactor terminals.

Button is attached to the hinge side of the door jamb with screws through the very thin brass flange. Contact is made or broken when button is depressed by closing of the door.

Entrance doors may be so directly situated above the concrete foundation wall that the hole cannot be drilled through the door frame straight down to the basement. An alternative that is usually successful is to drill a ¼-inch hole from the other edge of the door frame into the larger detector hole, then remove the shoe molding (the shallow molding atop the baseboard) and drill a hole about 6 or 8 inches from the door through into the basement. The wire is brought up from the basement, set into the wall corner covered by the molding, then rises along the end of the baseboard, and enters the hole in the door frame to the larger hole. If the shoe molding does not seat properly when replaced, just take some Spackle, applied by running the finger along the edge, and the molding will be sealed neatly and tightly.

In some construction, no space has been left between the door jamb and the studs, or the space has been plastered over. An alternative measure, then, is to remove the casing

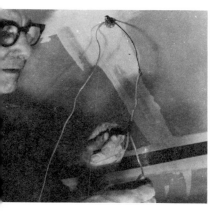

Rooms over the garage or an unfinished basement are easiest to reach, as wires can be brought below and carried along the ceiling beams to the new location. Wire is worked through ceiling panels of garage below the main floor. Here wire is taped to the end of the snake so that it can be drawn along the ceiling joists.

Panelboard ceiling makes necessary a few openings to pass wire around obstructing beams. Such openings in plasterboard are easily patched.

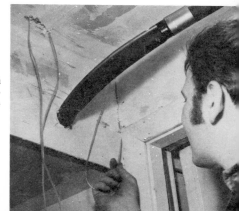

trim on that side of the door frame so that the wires can be recessed behind it. The casing is held in place with long finishing nails, and is easily removed without damage with a wide-blade knife or a flooring chisel. With the casing off, it will be easy to bring the wire up through a hole drilled straight down through the flooring, for connection to the detector button in the jamb. The wire is recessed into the space between jamb and stud. The casing can then be restored by replacing and countersetting the nails, and Spackling the nailheads.

If you hesitate to remove the door casing because it may damage or chip the wall paint along the edges, there's another alternative. This is to remove the door stop molding at the

Instead of tacking wire around a doorway, the molding trim can be removed without damage to reach the space inside.

Wire is placed inside the door frame between the jamb and wall stud, then door trim is replaced neatly.

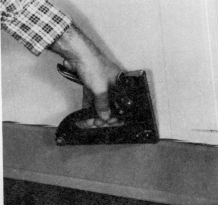

Wire is tacked along the top of a baseboard where it will be covered with the base shoe molding.

hinge side and drill a ½-inch hole under the stop molding position. The stop molding is farther away from the framing members so there always is an open space behind the jamb at that position. The wire can be fished through the second hole in the same way as described before, then a shallow groove is chiseled into the jamb between the two holes for recessing the wire so it is flush with the wood. The second hole is plaster patched, after the wire is drawn through. When the stop molding is replaced it will cover both the hole and the wire groove.

If the alarm system requires installation of an outside shunt lock and alarm indicator light, these usually are located on the latch side of the door jamb although the shunt lock may be located right on the door itself with a flexible cord. The installation usually involves the same wiring problems as the door detector button; the wires are brought up through the interior wall space, snaked in or by removing the casing trim on that side, as described before. There is one added detail, however. The shunt lock and indicator light are normally secured in position by tightening a lock nut on their barrel housing at the back, against the inside surface of the jamb. If space is lacking or inaccessible to accomplish this, however, an alternative method of installation is to drill the holes slightly undersize, then coat the barrel threads with epoxy glue and drive the tubular housings into the holes with hammer blows on a protective wood strip. The forced fit installation is done of course, only after the wires had been drawn up and soldered to the terminals of the shunt lock and indicator lamp.

One segment of door trim shown removed at lock side to permit installation of shunt lock switch and alarm indicator lamp into the side of jamb. Space between the jamb and framing stud makes it easy to draw up the wire.

Wiring Panic Button

In this series concealed wiring is demonstrated in wiring a panic button at a door. A practical way to carry concealed wiring from floor to floor, or between rooms and across doorways, is to place it under the threshold. The door saddles are usually held by just a few long finishing nails, can be lifted intact with a pinch bar.

Gap between flooring boards of the adjoining rooms provides space for tucking low voltage wires safely out of the way. Extension bit is used here to drill hole into an offset of the wall framing so wire can go through, rather than around on the outside.

Fished-through wire is recessed deeply behind the door jamb, in this case for installation of an unobtrusive panic button into the door frame.

Door casing is drilled with ⅝-inch wing bit for recessing body of the panic button, so it can be set in flush.

Small-diameter bit drills at an angle for hole to bring in wire at rear of the push button. The use of an extra long bit will help avoid damage to the door moulding from contact with the drill chuck.

Wire is easily drawn in from the side and out to the front and stripped.

Wires should be scraped always down to the bare metal just before attaching to the terminals, for trouble free contact. Make certain, also, that terminal screws are really tight.

Flush-type button is neatly pressed into the door frame surface, will hold firmly because of close fit. Wire along the side of the door frame can be held down with insulated staples or upholstery nails, completely out of sight.

207

Window Detector Wiring. Several types of detectors are used on windows. One push button type is recessed directly into the sill at the bottom, and functions by downward pressure of the window. An objection to this type of detector is that the sill is apt to receive the greatest amount of moisture which can affect the operation of the button.

Another type is fitted into the side jamb and its button is pressed inward as the window is lowered. The most dependable type is the completely enclosed magnetic switch, which has two parts — one is attached to the window frame and has the wire terminals; the other contains the magnet, and is attached directly to the sash. The magnet section operates the switch in the terminal section.

In these window installations, the wires can be brought up through a hole drilled in either the sill or the side jamb. In either case, it is necessary to drill through the sub sill and 2 x 4 supporting plates inside the wall, and also through the

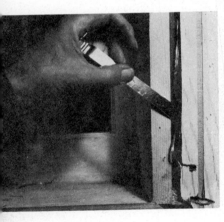

Where you can, remove molding trim at the side of window to bring in wires, which can be recessed into gap of the wall studding.

Hole drilled at angle brings the wire directly to position at sash where detector is to be installed. Minimum exposure of wire is desirable.

Window open, showing magnetic detector installed. The magnet part is on sash, the switch part is securely mounted on the window frame, turned so that the wire terminals are not accessible even if glass is broken.

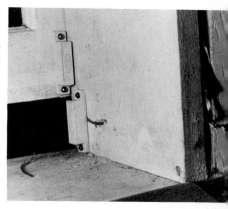

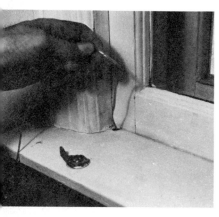

Wire can be dropped below window through hole drilled into the sill, brought down with a stiff wire lead.

Radiators under the window provide excellent means of concealing the wire installation. Wire brought down through the sill is carried through hole in floor board at radiator, or at clearance around the radiator pipe, to the basement or floor below.

floor plate. Depending on the height of the window, it is often possible to accomplish the opening with the 12" bit, drilling downward through the sill into the sub-sill framing, and drilling upward from the basement through the floorboard, to enable the steel snake to pass through.

Many windows are above a radiator, and you can use this situation to good advantage. Drill through the sill into the radiator enclosure, then try to snake the wire down around the pipe or tubing connected to the radiator. If that fails, just drill right through the floor boards inside the enclosed space to send the wire down. Usually, all the windows in a home are framed alike, and the effort to snake concealed detector wires at one window will show you the method for doing all the other windows quickly and efficiently.

Some windows are so situated that the wire tape cannot be snaked down through the floor. In those instances, an alternative method is to drill a hole in a diagonal direction just above the basement with an extension drill that will reach into the basement. The snake is passed down from the detector hole in the window jamb or sill and snared at the baseboard hole. A second snake is pushed down into the diagonal hole to the basement. By hooking the two snakes together, you will be able to pull the wires through all the way.

The small hole drilled above the baseboard is then patched with plaster Spackle and touched up as best you can to match the rest of the wall. The small size of the hole minimizes any damage to the decorations.

It is not good practice to run wires across floors under carpeting or linoleum; better to staple the wires at the baseboards around the room, either at the floor-wall corners, or just above the base shoe molding, although the latter is not entirely satisfactory. A helpful suggestion: remove the shoe molding, replace with a wider cove type that has a beveled back. This allows space for stapling the wires along the top of the baseboard. The wider surface of the new molding, when nailed in place, will also cover any chipping of the wall paint along the bottom edges.

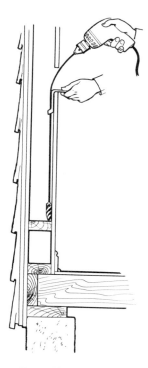

Flexible drill shaft, about 3 feet long, simplifies concealing wiring of window contactors. Drill can provide opening through the sill all the way through the floor boards to the basement.

Installing Heat Sensors. Fire alarm wiring is kept in an independent circuit to the control box if a separate signal, such as fire horn or siren, is used. The heat sensors and smoke detectors are installed on ceilings not closer than 3 feet to the room corner. This is a simple matter in one story homes in which the ceilings can be reached easily by way of the attic. The detectors are attached to the ceiling with screws, the wires carried through ¼-inch holes drilled in the ceiling and run along the joists. One difficulty that is usually encountered is caused by the insulation bats laid between the ceiling joists, which hinder passage of the wires down to the sensors. The problem is overcome neatly using a stiff piano wire to penetrate the insulation bat and carry the wire through. One ingenious mechanic has speeded up the process still further,

Wiring in Basement

Finished basement requires more care in concealing wires than unfinished. In this series a window detector is being wired. Basement ceiling molding is easily removed with small pinch bar so wires can be concealed behind molding. Molding is held with a few nails so it pulls off readily in short sections.

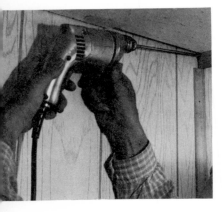

Long 12-inch drill goes through wall of basement room at ceiling corner.

Steel snake is fished through the hole from adjoining room. Wire, taped to the end of the snake, is pulled through.

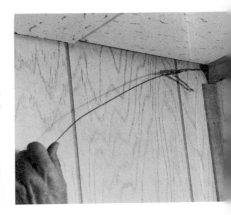

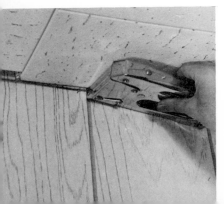

Wire is drawn along the ceiling corner, tacked down with rounded staples.

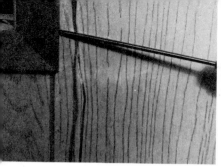

Entry through frame of window brings wire in from side.

Magnetic detector is installed at bottom of sash so very little wire is visible. Detector can be located at top of sash also. Magnetic part goes on sash; terminal part of detector is attached to the window frame.

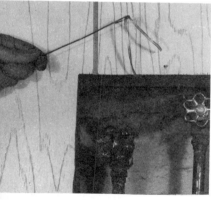

Every device and trick is utilized to keep the wire runs out of sight. Here a wall opening for access to water valves provides a means for guiding the wire snake through behind the wall.

using the drill bit that makes a hole in the ceiling to carry the wire through. The shank of the bit is smaller than its head, and has a small hole near the end. When the bit goes through the ceiling, it is left in position, the bit removed from its chuck, and a thin nylon cord tied into the hole. The wire is lashed to the cord. When the bit is pulled through at the ceiling, the cord and then the wire are drawn down

213

through the insulation, all in one stroke, without need for fishing through.

The heat sensors are usually located in interior rooms and at places different from the locations of burglar alarm detectors. Installation in two-story homes is done by reaching into ceiling areas from nearby closets, where some openings in the plaster must be made for snaking the wires to the required positions.

For example, a heat sensor in the kitchen would be wired from a nearby pantry or closet that is selected because the ceiling joists run in the same direction as the wiring to the sensor (or you are able to notch the plaster below any intervening joists between the closet and the sensor).

A hole is drilled in the ceiling for the sensor wire, a larger hole in the closet. The snake is put through the sensor hole and worked toward the closet where it is snared so the wires can be attached and pulled through. There is no difficulty bringing the circuit wires from the control box into the closet. Several fire sensors may be distributed to rooms in various directions from the single closet in that manner.

13

The Watchdog

If ever dogs had their day, this is it. In suburban and city homes, on lavish estates and in apartments, residents are coming to rely on trained watchdogs as friend and protector. Indeed, a loyal and alert pet dog may be the best protection your home can have—it is more sensitive than any alarm system, without the shortcomings that are inherent in any automatic device; it stands watch all day as well as at night, alerts you to the presence of a prowler before any attempt at entry is made, will hold an intruder at bay for a considerable period, and will leap to your defense if you are attacked. A dog is also a family pet and companion, and many people never go for a walk, day or evening, without the protective company of their dog.

The increase in daytime crimes at a time when women often are at home alone, points up even more importantly the 'round-the-clock protective role of the family dog. This dependence, though, can be placed only on an alert, sensitive, and loyal dog, preferably one that is trained to protection duties. An overaged, lethargic, indifferent pet just doesn't fit the bill. Let sleeping dogs lie! For safety's sake, you need one that's always on the job, eager to please and serve you, and the best way it can do that is to help keep your family and home safe by keen watchfulness.

While there are no statistics on the subject, wide-ranging conversations with homeowners and apartment residents, police administrators, and merchants, confirm the view that residences and business places which are protected by trained watchdogs or guard dogs are less subject to burglaries than similar places in their vicinities. In one specific instance, a small business place had been broken into time and again so that the insurance had been canceled, and all types of expensive protection including steel window gates and door alarms had failed to keep out the vandals and thieves. However, after the owner obtained a lively German Shepherd that had been aggressively trained for guard duty, all break-ins ceased and vandalism stopped completely. Many other store-keepers have taken to relying on dogs to protect their premises; owners of supermarkets and department stores found that the dogs have overcome one major problem by locating thieves who hide in the place just before closing, then simply let in their confederates to loot the place thoroughly.

In Los Angeles and other West Coast areas where crime levels are among the highest in the nation, commercial kennels report that the demand by nervous movie colony residents for trained watchdogs often exceeds the supply. John Wayne, whose valuable gun collection was stolen from his unprotected home while his family was asleep, has acquired several police dogs for his property. The Gregory Pecks were more fortunate, returning from the theater one evening midst tumultuous barking by their Wirehair Terriers. A private policeman in the area, alerted by the intense barking, had intercepted three masked burglars in the process of breaking into the Pecks' home.

Appeals to police by residents of Beverly Hills for better protection have brought the response: get a good watchdog, even a kennel of dogs, to keep tabs on prowlers. "If Joe E. Brown had owned a good German Shepherd at his place in

Brentwood, the robbery and pistol-whipping episode at his house probably never would have happened," said one private patrolman who covers that hillside area. He was referring to an early morning invasion by three masked gunmen of Brown's house at the time that the famed comedy actor was under round-the-clock care for a serious illness. The gunmen tied up two nurses, roused Mrs. Brown and bound her up also, then ransacked the house of $100,000 in furs and jewels.

These and even more serious recent events have prompted other inhabitants of the area to increase the number of private guards, at the same time shopping for large trained dogs that will add their full measure of protection.

In an outlying section of an Eastern city, where almost every home on several adjacent streets had been broken into within a period of a year, no attempt at entry had been made at one home that had a trained and very alert Alsatian. Subsequently, about half of the homeowners obtained similar watchdogs and the burglaries dropped almost to zero and remained so for almost three years. The sole exception occurred while the family was away on an extended trip and their dog had been left with friends.

Did the dogs keep the thieves away, or did prowlers simply shun the area because of the dogs? As with every other pro-

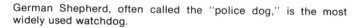

German Shepherd, often called the "police dog," is the most widely used watchdog.

tective barrier that they encounter, burglars have learned to deal even with trained watchdogs, using tranquilizer darts, chemical Mace, poisoned food, or muzzle traps. Well-trained dogs usually refuse food offered by anyone except their master, but chemically treated water or a dart could immobilize the animal. Luckily, few burglars are prepared to utilize these means of evading an alert watchdog.

What It Takes. Many questions arise when deciding on having a dog for home protection. These include, first of all, whether you can properly house the dog in your quarters and give it the humane attention that it requires. Beyond that, there is the selection of a breed which would be suitable for your needs and circumstances. There are also the questions whether you should raise a puppy or get a full-grown housebroken dog; what kind of training the dog should receive, and whether you can train it properly yourself.

Can You Keep a Dog? A dog, even one that is fairly large, doesn't take up much space and can adapt to almost any home situation. Still, there must be a suitable spot that can be assigned solely as the dog's own—for its food and sleep, and where it can get out of your way when it's not wanted around. Some people are unable to accept pets in the home; if that applies to you, go slowly. But things change, and you may have changed, too. The test, then, is whether you can accept the dog as a pet, a friend, and member of the household, without resentment and not merely as an instrument for your own purposes.

Don't count on Junior's present enthusiasm and promises to take all responsibility for care of the animal. While his sincerity is not to be doubted, it won't be long before school duties, sports and other activities, or seasonal trips to camp, will interfere, so that the care of the dog then rests with you or other members of the family. The dog, remember, must be walked at least twice a day—rain or shine, heat or snow—and occasionally taken to the park or out in an open field where it can be freed of the leash for a fast romp to get sufficient exercise. At least half an hour a day must be allotted for these outdoor airings. If such tasks would be onerous, better think twice before starting the routine. But remember that many people who had such attitudes and swore they'd never be tied to walking a leash become quite attached to the animal after they have it, and come to enjoy the walks in the open.

A Pal for Children. The presence of children in the family is an added reason for having a good watchdog, although it is essential to make sure that the animal that is selected is even-tempered and properly trained, to be certain that there would never be an injury to the child when your back was turned. Dogs quickly accept the children of the family, however, putting up with rough treatment from them that they would not tolerate from adults. There is another problem to watch, and that is allergies, but these usually cannot be determined in advance and you will just have to wait and see how it goes. If a known allergy to the dog's fur is discovered, there may be no alternative to disposing of the dog, or keeping it always outside or in the garage.

Going to the Dogs. Matters don't always go as they should, as pointed up in the following little item reported by Associated Press from the town of Romford in England:

"Ten dogs live on Turpin Avenue, and nine were in the doghouse today for having slept soundly as burglars looted 26 houses in the neighborhood, making off with a radio, some milk and biscuits, and about 100 British pounds in cash.

"Police said the intruders had worked the street for more than three hours before dawn without raising a single woof from a Great Dane, an Alsatian, a Boxer, and diverse canine friends. Finally, a 4-month-old Labrador set up a howl that drove the marauders away."

A dozen explanations may be offered to explain this novel story. Maybe English watchdogs are sympathetic to intruders who go for milk and biscuits for refreshments. Unlikely as the story is, it does point up the essential requirement of a watchdog: that it be always dependably alert. A tired, sleepy, indifferent dog does not fill the bill.

Selecting a Breed. Choice of a dog should be made with consideration for the ultimate full-grown size, temperament, adaptability, and capacity for learning. The German Shepherd, widely known as a "police dog," is the most favored breed as a protection dog, because of its intelligence, alertness, courage, loyalty, and other qualities, including size and aggressive bark. This a fairly large dog that needs plenty of food and exercise. But despite its size, it can be trained to behave when confined alone in a city apartment. Among the other popular large dogs that serve well for protective duty are the Alsatian, Doberman Pinscher, Dalmatian, Boxer, and Great Dane. Several medium size dogs, particularly the terriers, have plenty of pep and courage and make excellent watchdogs, adapt well to city life, and are fine pets. The favorite terrier breeds are the Scottish, Welsh, Kerry Blue, Irish, Airedale, Smooth and Wirehaired Fox Terriers, and the Bull Terrier. Some of the smaller breeds may serve your situation perfectly; among these are the Dachshunds, Poodles, and the Beagle. The yapping by an alert pip-squeak canine, remember, may be equal in effectiveness as a burglar alarm to the growling bark of a Shepherd. You don't have to scale the size of the dog to the spaciousness of your home, but it's worth giving this aspect your consideration. A German Shepherd would have plenty of scope in a house with a backyard or garden, while the confined space of an apartment would be better suited to a Dachshund or Fox Terrier.

Mongrel and crossbreed dogs, often available for the asking, may make perfectly satisfactory watchdogs. But there are some advantages in choosing a purebred dog, one of which is that you can be fairly sure what it will look like when full-grown, and also you can have some idea in advance whether a puppy has the capacity to be trained with the genetically derived attributes that establish behavior patterns in a specific direction. Some dogs are more capable of learning than others. Nevertheless, any puppy is an uncertain thing, since you can't tell how it will develop. For that reason, primarily, experts nearly always recommend starting with a dog 6 to 12 months old, so that you can judge its character and temperament.

Trial Period Important. Purchase of a trained dog must be approached with care. A trial period of about two weeks should be arranged, giving you an opportunity to see whether you can win the affection of the dog and also observing whether its behavior suits you. If you have decided favorably, there still remains one more step, that of having the animal examined by your own vet to be sure of its physical quality and condition. By this process, it may be necessary to try several dogs before making a final choice. Another approach would be to enlist the services of a professional security dog trainer and have him select a young dog, perhaps 10 to 12 months old, and arrange for the training program which you attend also to establish a "team spirit" relationship with the dog, who is being trained to protect you while you are being instructed on how to handle him.

Buying a Dog. Prices for full-bred, housebroken but untrained watchdogs generally run from $50 to $150, and up to $500 for some breeds. Whichever you select, be sure to obtain a written guarantee from the kennel, and an option for an exchange for any reason during a trial period of at least two weeks, just in case you find that you don't care for that dog after all, or it shows that it doesn't care for you.

As to the choice between male and female, there are pros and cons for both, but some trainers favor the female as more obedient, willing to learn, and with a greater protective instinct. Always watch out for the animal with a mean streak,

Doberman Pinscher is known as fierce, but a dog's nature depends much on training.

Dalmation, the "coach dog," makes a good pet as well as a watchdog.

The Boxer is a strong and energetic watchdog with a fierce appearance.

The lovable little Dachshund makes a good apartment watchdog.

The Bulldog's appearance is enough to frighten a prowler.

The beautiful Weimeraner has been successfully trained in recent years as a watchdog.

The Airedale Terrier is a aggressive moderate-size dog.

The mammoth Great Dane may present a feeding problem.

The Rottweiler is increasing in popularity as a watchdog.

the one that shows a tendency to break out of the trained pattern of behavior and make a vicious, unprovoked attack.

The basic types of training for protection services are: 1. guard dogs, which receive just minimum training and are used to maintain surveillance over a specific property area. They are of an unfriendly and uncertain nature, and are generally left on their own in a confined area; 2. security or patrol dogs, who work with handlers and are trained to discover hidden persons in public places, such as park or company grounds, department stores, and industrial plants after closing hours. The patrol dog is finely conditioned and responds directly to the commands of his handler. It will attack and seize a prowler when ordered to do so, but is otherwise docile and obedient; 3. attack dogs, lethally aggressive, are often used by the military to ferret out infiltrators, and by prison guards patroling open compounds. An attack dog is so vicious that it can be controlled by only one person, and usually works with a single master. Such a dog can be expected to attack any person who comes near unless restrained, and for that reason cannot be allowed in civilian areas. As one dealer described it: "An attack dog in a home would be as lethal as a loaded pistol on the kitchen table."

None of these training programs fits the needs of civilian owners, certainly not those of family life where a dog must obey and protect several masters, learn to accept many people who are received by the household, submit to the teasing of children, and generally behave decently, all the while remaining alert to the out-of-the ordinary, the threatening danger. The fact is that the properly trained, loyal dog can do all this, whether it comes from instinct or, as some believe, a difference in smell from a source of danger, or a certain intelligence. People can come and go in a house, some of them strangers, and the dog will remain unconcerned. But at a canvasser ringing the bell, or the approach of some person with a "different" appearance in clothing or demeanor, the dog will spring up rumbling and ready for action at the go-ahead word.

Owner's Liability. Always remember that you are responsible for the actions of your dog. If you choose a menacing, uncontrollable animal and it terrifies or inflicts injuries on an innocent person, such as a postman, neighbor, or visitor, you can be sued for negligence. Your liability insurance policy probably covers this situation, but don't be surprised if your policy is quickly canceled. Canvassers have a legal right to come to your door without being attacked by a dog. And don't think you can sic your dog with impunity on anyone who strolls into your backyard by error, or on boys who leap your fence to pick fruit or flowers, even though they admittedly are trespassing. Unless they were committing a serious crime, the most you can do is chase them off the property. If you have a vicious dog which is allowed free rein of your yard, an injury that it causes to unsuspecting persons may be your responsibility. Just posting a warning sign at possible points of entry is not always regarded as sufficient notice. The dog must be restricted to a separate area. A restraining chain on an overhead runway wire may be the most effective way to meet all the requirements, allowing the dog plenty of leeway for exercise and to protect your property, while keeping it within reasonable bounds.

Training Your Dog. Any alert and loyal dog can be depended upon to some degree for protective services. Their effectiveness increases immeasurably when they become trained to obey commands, responding only to their master and those who have been accorded the approval of the master. Obedience training also protects the dog from injury by passing cars, eliminates annoying habits such as snatching food from the family table or jumping up on visitors, helps adjust the dog to living in the civilized human environment which requires it to quietly accept long periods of being left alone, and generally makes the animal a more congenial and understanding companion. A dog that has not learned to behave would become a nuisance and a menace, and cannot remain for long in a city home.

Obedience training calls for patience, determination, consistent efforts and at least rudimentary knowledge of training techniques. The essence of canine training is firm control by the master over the dog. This control is achieved through conditioning the animal to respond, without hesitating and almost by reflex action, to certain signals or stimuli. Experts claim that the reason few people can properly train a dog is, mostly, because they don't take the trouble to acquire the necessary knowledge, also because they fail to patiently devote the amount of time that is necessary to accomplish the desired result. More than that, owners permit bad habits to develop, then deal improperly at best with conditions when it may be too late to correct them.

Obedience training can be started when the dog is six months old, but not any earlier than that. The degree of success you achieve will depend largely on your consistency, in both commands and attitude. During the training period, commands are to be given always with the identical words and tone. Avoid even minor variations in the sequence in which the words are used. The training sessions should lead to a happy rapport between you and your dog, which means that you are never to lose your temper and resort to angry shouting or nagging repetition. Control over the dog will flow readily when you have full control over yourself. What you fail to accomplish one day may well come easily the next session.

The words of command should be selected carefully. They need to be short, each one phonetically distinctive. You may pick your own, or use these standard words:

SIT	STAND	NO	KENNEL
HAND	UP	STAY	AWAY
HEEL	FETCH	CARRY	WATCH HIM
DOWN	JUMP	QUIET	STOP

Training procedures, and the various tricks that can be used to impress a certain conduct on the dog, are too detailed to be elaborated here. Useful instruction on training can be obtained from several manuals on the subject, excellent examples of which are *Obedience and Security Training for Dogs* by Tom Scott (Arco Publishing Company, New York), *The Complete Book of Dog Training and Care*, revised edition 1962, by J. J. McCoy (Coward-McCann, Inc., New York), and *How to Train Your Dog* by Ernest H. Hart (T. F. H. Publishers, Jersey City, New Jersey).

Classes in dog obedience are conducted by the American Society for the Prevention of Cruelty to Animals (ASPCA). More than 25,000 owners and their pets have benefited from this instruction since the program was started in 1955, and usually there is a waiting list for admission to the classes. Films and slides are used in an audio-visual center in conjunction with actual training demonstrations. The program has as its objective the control of unruly dogs, making them both more acceptable to people generally, and helping the dogs adapt better to the surroundings. Information about these classes may be obtained by writing to the ASPCA headquarters at 441 East 92 Street, New York, N.Y. 10028.

If the services of a professional trainer are obtained, it would be wise to accept the trainer's judgment as to the character and learning capability of the dog. Do not persist just because you've become attached to the animal as a pet. If the animal is deficient in intelligence, hearing, eyesight, stability of temperament, or the kind of alertness that shows a devotion to its master, you would do well to dispose of it at once, possibly to a friend who wants merely a pet. The trainer very likely will be willing to seek a substitute for you that will meet your protection needs—a dog that can be trained to obey and will stand guard over your home night and day.

When the dog is being trained, it is important that you attend and become part of each session's activities, to establish a team relationship with the dog. If attendance is not possible, then experts advise that the dog be kept by the owner for several months before it is given for training, or that the training program be conducted right in the home.

Cost Adds Up. Everyday maintenance costs, mostly for food, are not considerable for a small or even medium-size watchdog, but feeding an Alsatian or other large dog can run up a major bill. Other costs are veterinary charges, which can be substantial, and outlays for supplies, boarding in kennels while you're on vacation trips, license fees, and tips to the doorman or fees to student "dog walkers" for times when you can't get home to take the dog out yourself.

Explanation of Rabies. An estimated 600,000 persons are bitten severely enough by animals each year to require medical attention. Unless the attacking animal is known, or can be identified and watched for symptoms, the threat of rabies is present and the victim must undergo a series of 14 painful injections to stave off the disease. Rabies is an acute disease of the central nervous system to which all warm-blooded animals, including man, are susceptible. The virus of the disease is present in the saliva of an infected animal, and is transmitted by bites, or licks of raw skin surface. The incubation period varies from 10 days to over a year; once the disease has passed beyond the incubation stage—and symptoms become apparent—rabies invariably causes death. There is no effective treatment, only protective measures to prevent its development.

A perfectly healthy dog may become rabid if bitten or licked by an infected dog or any wild animal—this is most likely to occur at times when your dog is having a free run in open fields. All bite wounds, and any abraded skin area that had been subject to the licks of suspected animals, should be washed immediately with soap or a detergent solution, making sure to reach into deep puncture wounds. Anti-rabies serum and vaccine should be given if the attacking dog showed signs of having rabies or the wounds were inflicted by a wild animal.

The early indications of rabies in dogs, often so slight that only a trained observer may note them, include fever, failure to eat, excessive sensitivity to sound, to touch, altered disposition, and frequently, change in the tone of bark. The excitation phase occurs within a few days, marked by abnormal restlessness, erratic and aggressive behavior with snapping at other animals and humans. Death usually occurs within 10 days.

Apartment Problems. Watchdogs aren't always welcome in apartment houses. Landlords insist that dogs cause structural damage, particularly to the woodwork, and litter the halls and elevators. Some tenants are terrified by dogs, complain about their barking or movements in apartments. Many leases contain bans on pets of any kind, and ownership of a dog may result in eviction. Even cooperative and condominium apartments, in which the occupant has partial ownership, have restrictive rules regarding dogs. Owners of such apartments have on occasion gone to court to seek easing of these restrictions on the ground that the dogs were needed for protection, citing numerous crime incidents in the building or neighborhood. In rare instances a friendly or sympathetic judge has ruled in their favor; most other suits of this kind have resulted in findings by the courts that the need of a dog for protection did not justify violation of an occupancy agreement prohibiting maintaining a dog in the apartment. In one decision that was favorable to the tenant, the judge stated that "this court is not prepared to hold unquestionably that a prohibition against animals need be unyieldingly enforced at all times, particularly where the occupant of a cooperative apartment has shown a series of burglaries in the past few months." This ruling was overturned on appeal, but in other such cases the tenants have succeeded in confirming the right to keep a pet for protection.

Many building owners have become more lenient on the issue of keeping dogs, and in fact they've come also to depend on the protection of an alert watchdog for their own homes. This movement is being pushed hard now by tenants who no longer are willing to simply abide by whatever regulations are forced upon them, prejudiced as they may be, without regard to their own rights and needs. In addition to individual court actions, groups of tenants in larger apartment complexes are signing petitions to obtain relaxation of the no-dog rule. "We have been told that we must get rid of our dogs. We, the parents, and our children are very emotionally upset about this," said a petition signed by 13 tenants of a 40-family building. "Aside from the love we have for the dogs, the families in our building feel more secure because of the watchdogs." The result of this and similar petitions is that quite a number of apartment owners have relented, granting permission but specifying conditions under which the dogs are to be maintained.

In those apartments where keeping a dog is permitted, correct behavior of the dog becomes the critical requirement.

Since the dog will necessarily be confined much of the time in a limited area, the breed, character, training, and care of the animal are all important. Size is not necessarily the decisive factor, since certain large dogs often fit in well with apartment life, needing only an occasional free run in open country. What is essential is that the dog be trained to wait patiently for his master's return, when it will be taken outside. A restless dog, one that bounds around the apartment when alone, can create considerable disturbance in the building and soon will be the subject of complaints. Barking is usually better accepted if the neighbors are friendly, though don't expect them to remain on agreeable terms if you insist on maintaining a hound that howls and moans constantly. A noisy dog will soon find itself out on the street—with his master for company.

Training a dog to be quiet in the house is quite difficult, and doubly so since you want it to be free to raise an alarm when danger threatens. The key word is "Quiet," and it is impressed upon the dog by every possible means. Owners have even used slingshots to get the message across. More effective, though, is attention at reasonable time spans. An intelligent dog will quickly learn that its master will arrive to feed him and take him for a walk, and therefore the impatience will subside. This attitude cannot be developed if the arrivals are irregular, or occasionally skipped altogether.

14

When You're on Vacation

Nothing can spoil a vacation so completely as having your money and luggage stolen while traveling, or on your return finding that the house has been broken into and thoroughly ransacked. Worrying about these possibilities won't solve them, and you certainly should not give up the trip for that reason. Go, of course. But a little extra planning before you leave may prevent these disasters and allow you to have a good time.

Ideas and suggestions are presented here that may strike you as effective methods for keeping your home safe from prowlers, and your holiday excursion undisturbed. A checklist of the standard rules, even though often repeated, is given here as a reminder:

1. Stop deliveries of milk, newspapers, bread, and similar services "until further notice," (but don't explain that you'll be away on vacation).

2. Have a friend or neighbor pick up your mail daily, or ask the Post Office to hold the mail until called for. Better, perhaps, is to rent a postal box for the time you'll be away.
3. Arrange for lawn care and general outside cleaning, including removal of advertising circulars left outside your home.
4. Make provision for lights both inside and outside, turned on and off by time clocks or light-sensitive switches.
5. Leave a key with a friend or relative who can check the house daily to make sure all is in order—lights working, heater on if the weather turns cold, etc. Authorize him to call a serviceman in an emergency.
6. Notify police of your vacation plans, so that the house can be given extra patrol attention. Leave an address where you or a relative can be reached. (Be sure to let the police know when you return so you won't be mistaken for an intruder.)
7. Don't let your trip be publicized in the press or announced generally around the local stores.
8. When leaving, barricade all doors (except the one you'll use) with suitable lengths of 2 x 4s as extra braces to forestall jimmying.
9. Put all valuables out of the way, preferably in a safe deposit vault or a security room safe. Keep a separate list identifying serial numbers of valuable items. (Write to the Schlage Lock Company for a handy card arranged for listing identifying data such as model and serial numbers.)
10. Don't forget to lock the garage doors. Any ladders that are accessible should be securely chained and padlocked.
11. Leave the house looking as it usually does; avoid the "closed down" appearance such as drawing all the draperies and pulling shades all the way down.
12. Lawn sprinklers are just dandy—set them up with a time clock to go on for short intervals each day. The impulse-type sprinkler also can be clock controlled.
13. The garbage cans will be unused—best to place them out of sight so they won't confirm any notions about the

house occupancy. Keep them in the garage if there is space.

14. Suspend your telephone service — better to have a "temporarily discontinued" announcement than a repeated "no answer." If the phone is kept in service, cut the bell down to softest tone so continued ringing won't be heard outside.

15. Test your burglar and fire alarms, and be sure to turn them on when you leave. But make provisions with a neighbor for entry in the event of a false alarm that must be turned off and give him the number of a serviceman.

Home Sitters — The New Wrinkle. All the preceding rules can be handled routinely in preparation for your vacation trip. There is still a further protective measure, if you have a particularly valuable or vulnerable house, or are concerned about the safety of an art collection or other treasures, or you have a pet that you'd prefer not placing in a kennel. The answer to all of the above may be to obtain the services of a temporary house sitter.

That means having a really dependable person who will live in the house while you're away, take care of the animal, water the plants, collect the mail, take any necessary action in the event of an emergency requiring a plumber or electrician, for example. Having a trustworthy person occupy the house while you are on a trip is perhaps the best assurance of avoiding a sad experience.

How to find a suitable home sitter? If you live in a college town, you'll no doubt obtain the services of a reliable student at very modest cost. The school registry office will gladly assist you and make a recommendation. Sometimes a local industrial plant can recommend a person who may have recently relocated and not yet settled down in new quarters, so would be happy to have such a temporary assignment. An elderly person living with relatives may be the most suitable candidate, one who can assume responsibility and understands the requirements of keeping a home protected. Of course, you will check on the character and living habits of the person you select before making a firm commitment.

Lights On! One of the best ways to have your home avoid the "family out of town" look is to keep the lights on. A single light inside the house, however, no longer is sufficient—it may even serve to give an intended intruder confirmation of his observations. In fact, the house with a single light burning constantly in the same upstairs room, night after night, may even attract the attention of an observant burglar cruising the area to mark his targets. Aware that people just don't live that way, he'll reach the conclusion that the solitary light is just a blind; he'll consider it almost a beacon light.

Lighting, even with costs of everything constantly soaring, still is quite inexpensive. A 100-watt bulb can be kept burning all night for only 2 cents. Suppose you had two 60-watt bulbs, each in a different room, turned on and off at different times by an ordinary timer clock or photoelectric switch? The expense would be nominal, about 15 cents a week or $7.50 a year—a worthwhile outlay for this effective security measure that may save you many times that amount. An automatic Technitimer, to turn lights on and off at specific intervals, is available from Lafayette Radio for $7.95.

Many knowledgeable authorities suggest also that the sound of a radio helps give the impression of someone at home. You can easily plug a small radio through the timer clock to go on and off at certain intervals—not necessarily throughout the night. The air conditioner also can serve a similar purpose—just the fan (not the compressor) working would be sufficient.

Outdoor floodlamps, turned on automatically every night, are of particular benefit in guarding the home while you are away.

Pattern of Light. The traveling family could feel greater assurance if the home were equipped with the sophisticated light pattern controller first presented in the June 1971 issue of *Popular Science* magazine. This is an electronic unit that controls the lights in six rooms, producing 11 varying patterns of lighting effects which would dispel any prowler's notion that the family might be away.

The "people simulator" device created by Ronald Benrey

literally duplicates the family activity by continuously shifting the pattern of lighting throughout the house, turning lamps on in some rooms, while others go off at various times. Before dinner, lights are on all over, during dinner only the kitchen and dining rooms are lighted, later only the family room has lights for the television viewers, then lights appear in all the upstairs rooms at random times, and also in the basement workshop. Late at night, all lights go out except for just one or two which remain on until morning.

The system is based on an interesting electronic concept. There is no change in the house electrical wiring—instead, in each room, a lamp is plugged into a standard 3-prong grounding type receptacle which has been converted to serve as the switching medium, and is in turn connected by a cord and plug into the regular grounded house receptacle. The switching receptacle is encased in a standard metal electrical box for safety.

A single thin wire for low-voltage current connects this switching receptacle in each room to a central controller, which is made up of an ordinary electric timer (such as the Lafayette Radio Intermatic No. 13K-27022 at $9.72) to which a snap-action switch has been added. Metal spacers, cemented to the timer's rotating dial at specific time-interval distances, serve as the trippers that snap the switch at various times to send low-voltage pulsations to a transistor. This in turn operates a 12-position rotary switch which supplies current to a diode, of which there are 11. These diodes control the distant lamp outlets. In this system, while there are only six lamps in as many rooms, these can be managed to create 11 different lighted patterns. The 12th activator trips the rotary switch to "off."

A few precautions are suggested for safe and reliable operation of the lighting system. The outlet supplying current to the controller must be the three-prong grounding type that is correctly linked to the house ground; the circuit ground and metal case also must be connected to the house electrical ground. Polarity must be preserved in all wiring so that the system conforms with the National Electrical Code. If you follow the diagram carefully the pattern lighting "people simu-

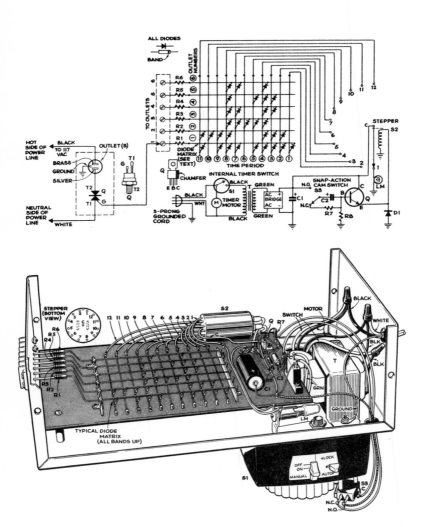

Robert Benrey's "people simulator" is a good project for the electronic hobbyist; it requires a little more savvy to build than most home-protection devices.

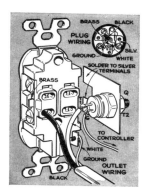

PARTS LIST

C1—1.000-mfd, 15v capacitor
C2—150-mfd, 15v capacitor
R1-R6—100-ohm, ½w carbon resistor
R7—100-ohm, ½w carbon resistor
R8—560-ohm, ½w carbon resistor
D1—All diodes 1N914 (see text)
Bridge—Bridge rectifier (International Rectifier 50FB05L)
Q—GE D40D1 transistor
Triac—All triacs GE SC40B (see text)
T—6.3v @ 3A (Stancor P-6466)
LM—No. 47 pilot lamp
S1—in timer (see text)
S2—Stepper switch (Guardian IR-705-12P-6D)
S3—Snap-action switch (Micro VLD-0017)
Misc: 24-hour timer, grounded cord set, 12" x 7" x 4" aluminum minibox, wiring boxes, double outlets, spacers, copper strip, twist-on connectors, #18 solid wire, #20 hook-up wire

lators" can well be left to function on its own while you go off and have a good time.

On Your Way. With the house properly secured, you can start your vacation with a carefree frame of mind. But there are still a few details that should be gotten out of the way. You've listed the traveler check numbers for replacement if lost or stolen, but did you also write down the regional addresses of the agency issuing the checks so you can make quick contact if necessary? And did you weed out your credit cards, down to the one or two or three at most, that would likely be used on the trip?

Now, where do you keep your wallet and important papers, such as the passports, traveler checks, plane tickets? Not in a bag that's checked through on a plane flight. Inside coat pockets with zippered flaps, or special money and document belts, have saved many a traveler from the great distress of being stranded without cash or identifying papers.

Baggage Insurance. Before leaving, look over your homeowner's insurance policies. First check on the coverage for losses away from home—you probably will find that you're covered for lost or stolen luggage and clothing to a stated amount, usually up to $1,000. This may save the cost of taking out separate baggage insurance. If you feel that the amount is not adequate, buy separate baggage insurance from your travel agent or insurance broker—the cost runs about $15 for $2,000 coverage, but this new policy replaces rather than supplements the homeowner policy, as you can't collect from both for the same loss. Better still, take along less baggage—you won't miss it.

While you are checking the policies, keep in mind any vacation activities that may require special insurance protection. For example, will you be renting and transporting a boat or camping trailer to various lakes? Look to see whether the policy excludes such rented equipment—if so make sure when you rent to obtain the necessary insurance protection as well, covering the rental period.

All insurance policies specify a low limit on recovery for loss of cash. Your protection is to carry travel checks, which are guaranteed for replacement of any that are lost or stolen.

Make sure, too, that your homeowner's policy is in force, paid up to date and through the period that you will be away!

Safe Baggage. Insured or not, loss of any part of your luggage can put a crimp into your vacation. Lost baggage very often results from mistaken identity—so many suitcases look exactly alike at quick glance to the hurried traveler, and porters also are apt to cause mixups at airports and hotel lobbies. Your best protection is to place an identifying mark on all your luggage such as a distinctive stripe of colored masking tape along both sides of each article that you have taken along. Also, be sure to include identifying tags with your address.

One way to make certain that your luggage stays with you through the trip is to keep your eye on it during transit changeovers, particularly at airports. Directly after arriving on an incoming flight, go to the baggage department without delay to claim the luggage.

While often there is a considerable wait, at times the luggage comes through with unexpected speed, and yours may disappear—by error or intent—if you wait too long. Although the airlines try to check on pickups by individuals, honest mistakes occur frequently, and it is still notoriously easy for thieves to take items off baggage carousels and simply walk

Door-jamb is handy and effective protection for travelers, slips easily under doorknob to block entry even with the key. The unit extends to required length, telescopes small enough to fit into suitcase.

off in the crowd. At all transfer points, it's a good idea to stand by and count your luggage as it arrives or is being loaded into bus or taxi.

Room Safety. There are simple protective measures that can assure your safety in hotel and motel rooms. One of the most effective and convenient devices is a lightweight tubular post, forked at one end to fit snug under the door knob, and bent sharply at the other end so its rubber tip grips the floor, preventing the door from being opened even by a person having a master key. The post can be adjusted in length by slipping a pin into variously spaced holes in the tube, but as nearly all knobs are at the same height, adjustment will rarely be necessary. However, the variable length permits fitting the post into a suitcase for transport. Because it is so easy to set into place the post will be used regularly, while some other devices are so difficult to place that the tendency is to neglect them.

Various types of this door guard have been developed. The one with a forked top, called Protect-O-Post, is available from any Gimbels store for $8.95. A modified type, with a large slot at the end to fit under the doorknob shaft, is called Door

239

Jam, and is sold by Hammacher Schlemmer. The travel model, which telescopes to 18-inch length, sells for $5.00.

Door Alarms. Supplementary locks, some armed with battery powered siren alarms, are useful when traveling in areas where hotel security is doubtful.

The "Secur" intruder alarm is designed for convenient use on hotel and motel doors. Compact and portable, the alarm unit is suspended from an inside doorknob, and is plugged into a nearby 110-volt receptacle. A shrill buzzing noise is sounded if anyone touches the doorknob or forces the door without touching the knob. An on-off switch is illuminated on a dial. Available from Allied Radio and Electronics, price $19.95.

An inexpensive burglar alarm, the Sentinel, has a metal plate that snaps over the top of a door at the hinge side, so

Ultrasonic alarm described earlier is an easily portable protection device for traveling. It detects any intrusion regardless of the means of entrance.

that opening the door releases a projecting switch lever. Flashlight batteries power a siren horn. Distributed by Puretec, Inc., and available at housewares and hardware stores, under $5.00

Fire Alarms. Whether or not in doubt about the fire warning arrangements at the various hotels you encounter on a long trip, you'll feel more comfortable having your own fire detection device along, placed in a strategic spot high up in the room. These are, of necessity, small, battery powered units operated by a fusible link which may or may not be precisely accurate, and usually are not adequate for home installations. The obvious shortcoming of these units in the hotel environment is that they would function only to report a fire that is actually in the room, while the peril, of course, could come from combustion gases originating in other areas of the building. Simple, flashlight-battery alarms are priced as low as $3.00

More functional and dependable than an ordinary fire alarm is a portable smoke and heat detector that simply plugs into any 110-volt receptacle. The unit is completely self-contained, detects presence of smoke by means of a photo-electric cell. From Lee Electric Co., price $71.00.

15

Protect Your Car

Although more car owners are installing anti-theft alarms and devices, the number of thefts is still soaring. In 1970, more than 915,000 cars were stolen, an increase of 173 per cent over the past decade.

If you live in or frequently drive into a city, the chances are one in forty that your car will be stolen. In the suburbs and rural areas, the rate of car thefts is much lower.

While approximately 84 per cent of the stolen cars are recovered, the damage and incidental costs run the loss up to almost a billion dollars. This loss is suffered not by the owners of the affected cars alone, but rather by every insured car owner, because premium rates reflect the total cost. Thus each owner, no matter how careful, shares the burden and is penalized for the prevalence of this crime.

Where the theft victim individually suffers most, aside from the money loss, is the endless inconvenience that results from having a car stolen. The theft may occur under conditions that leave the owner stranded in a strange or outlying section. After renting a temporary car, there are the innumerable police reports in addition to filling out an insurance claim, obtaining replacement license plates, and also notifying oil company and credit card companies about any charge cards that had been left in the car pocket—and this latter action can mean temporarily blocking use of all duplicate credit cards with the same companies.

After that comes the inevitable hassle with the insurance company. The claim settlement is based on book value, which is what a dealer would pay for the car based strictly on model year and equipment—through your car may have had very low mileage—and you face an enormous outlay for a new car. Or if your car is recovered and returned to you after being used for wild joyrides, the abuse it received may not be immediately apparent for basing a claim for the damage. Perhaps even more distressing is the peculiar repercussion in which the insurance companies react by cancelling or limiting the policy—in effect punishing the victim for the crime.

Most are Youths. Studies by the F.B.I., the National Auto Theft Bureau, and other organizations show that approximately 70 per cent of car thefts are for "joy riding." Kids steal a car for an evening or two of hell-raising, and then abandon it. These offenders are nearly all under 21 years of age, and more than half are less than 18 years old. Such thefts generally result when the opportunity arises—the young offender grabs a car on the spur of the moment, just because the opportunity presented itself—a car left conveniently ready to drive away, doors unlocked and motor running, or keys in the ignition. But not always is the fault with the owner; quite often teen hoodlums roaming the streets decide to heist a car, and pick on the most likely one around.

They may not have all the equipment, but they know nearly all the tricks about opening locked doors, and jumping the ignition and starter solenoid. But it's obvious that the well-protected car will be passed up for the easiest one, and these youngsters often simply try the doors of cars along the street or parked in a lot until they find the one that seems to say "go." And away they go!

More difficult to balk are the expert car thieves, the "pros," who account for almost one third of the total thefts, according to the National Auto Theft Bureau. The targets are nearly always popular late model or very expensive sport cars. These "pros" not only know every trick of defeating car alarms and antitheft devices, but also often have the right key.

Filling an Order. In a gang-organized operation, a car that is spotted as the right one to fill an "order" is checked out for serial number while the owner may be in a restaurant or at the theater. Usually the door lock is pulled out bodily with a "bam-bam" device. In a short time, the right master key is delivered to the gang member at the location, or the key may be made right at the spot with a portable key cutter according to the code book. If the master key is not available, the gang operators have been known to obtain access to the restaurant's cloak room, examine the victim's keys, and simply make a quick duplicate from either the code number or a quick impression in sealing wax. The rest is a matter of simply using the key to drive away. The serial numbers are changed, the car given a new identification, and soon has a new owner.

Many of the stolen cars are immediately driven out of the state, or shipped out of the country. And more frequently than one could believe, almost-new cars are stolen just to be dismantled for their parts. A roundup of 17 persons in a car theft ring revealed that 15 to 20 late model cars were stolen every week to be dismantled for their parts. The ring was accused of stealing more than 1,000 cars "on order."

Stripping—A Vicious Vice. Many an owner who can't start his car soon learns the reason—the battery is gone. Or maybe he finds his car jacked up and minus one or two wheels, because someone had a flat and needed the tires, or some young delinquents were offered a couple of dollars for tires of that size. Even whole radiators· have been removed from cars right in front of their owner's home. Locked trunks of new cars are forced open with a pinch bar and the spare tires swiped.

These acts have stimulated counter efforts. Car hoods are

controlled by underdash locks, combined with alarm sirens that sound off if force is attempted. Car trunks also are included in the anti-theft alarms. Radio antennae which vandals took some kind of pleasure in snapping off the cars, now are virtually eliminated, replaced by wired windshield glass. Door handles are recessed, no longer easily gripped and manipulated by a wire loop. Even the windshield wipers are recessed on some cars to prevent their theft.

Making Your Car Harder to Steal. The most effective theft deterrent is "always to lock the ignition, make certain all the windows are rolled up, and all doors locked every time you park," even for a few minutes.

It may seem like a nuisance to do this, but it's nothing like the inconvenience you'll face if a youngster bounds into the car and takes off. Electrically operated windows, and door locks that can be set from outside without a key, make it easy to follow this protective rule. If you're concerned about locking yourself out of the car, keep an extra key tied with wire in a secret place that can't be spotted easily. Avoid the usual habit of placing the key in a magnetic case inside the bumper rail—that's almost the same as keeping a house key under the door mat, as every youngster knows all about that favorite spot, and it takes him only a minute to locate it.

Anti-Theft Devices. Many types of alarms and protective devices are available. Some are intended to prevent starting the motor, or turning the steering wheel, or shifting the gears. Others sound a loud alarm, by siren or shrill horn, to scare the thief or attract police attention, when a door, hood or trunk is opened. Another type of alarm is a contact sensor, which switches on lights or sounds a siren if any part of the car is touched by a person standing on the ground unless the proper door key is inserted. Ingenious electronic devices have been developed that require a coded card or number sequence to start the car instead of a key, and another clever installation has a time delay switch which goes into action soon after the car is started and under way—then everything goes dead. But simple secret switches seem about as effective as any other means of theft deterrence.

An interesting device developed by a West Coast doctor, who had been victimized by car thieves, is the Burgle-Bungle, which utilizes a solenoid to control both the hood latch and ignition system. The hood remains bolted shut, to protect the battery against theft, while the ignition is completely cut off by a secret switch.

Triple Control Lock. One of the most hopeful developments was the innovation by General Motors of a system that with one turn of a key locks the ignition, steering column, and gear shift lever, and the key cannot be removed until the gear lever is shifted into "park," thus putting the car into a braked situation as well. Additionally, a buzzer sounds to remind the driver if the door is opened while the key is left in the lock.

Introduced in the 1969 models, the combination locking system shaped up well in the year-end theft statistics. While auto thefts generally continued to rise, the National Automobile Theft Bureau reported that thefts of G.M. cars had dropped 23 per cent in that year.

That excellent and hopeful result still must be tested over a longer period. Two shortcomings have shown up, however, since the locking system became standard equipment on the 1970 G.M. cars. One is that the inability to remove the key without shutting off the motor during short stops may discourage locking the doors; a driver who stops to run in "just for a moment" to pick up the newspaper or on a similar errand leaves the motor running, and doors unlocked, to return and find that his car has disappeared. He's made it too appealing for some roaming juvenile to resist.

The other deficiency is similar to that of previous car locks — the keys are easily obtained. The G.M. locks have 1,000 different combinations, so chances of duplication are remote, but the master keys are easily obtained by the "pro" who has access to code books and similar sources. Master keys are traded in the underworld black market, giving even the novice crook an open access to your expensive car. It's not an easy problem to solve, certainly.

This would seem to make a discouraging picture. Michael J. Murphy, president of the National Auto Theft Bureau and a former Police Commissioner of New York City, has stated that so desirable a device as the G.M. door buzzer does not have complete consumer acceptance, and that many drivers, annoyed by the noise, simply disable the buzzer.

Types of Alarms. Here are examples of other vehicle alarms, their sources and approximate prices. This is not intended as a complete list, but rather as a guide for selecting the most suitable type.

A complete siren alarm system, with switches for the doors, hood and trunk (in two-door cars the extra switches may be used for protecting a tape player and attaching to the emergency brake), a key-lock switch that mounts on fender or cowl to turn alarm on and off. Kit includes a relay, siren, all necessary hardware, and wire. Available from J. C. Whitney & Co., Price $24.89.

Time-delay alarm protecting trunk, hood and doors. Fender need not be drilled for key switch; a thermal relay allows 7 seconds after turning on switch to leave the car, and same time lapse when entering. Horn mounted under hood sounds off if secret switch is not disconnected in time. Also from J. C. Whitney, price $15.88

Police-type alarm siren, for cars, boats, campers, 12-volts only. Includes 4 door switches, hood and trunk switches, key lock, siren with relay, all brackets and wire. From Universal Security Instruments, Inc. Price $25. Siren alone $14.50.

Hood lock, protects battery and prevents jumping ignition wires. Electric type, connects to ignition switch, cannot be operated until key is turned to "on" or accessory position, Price $8.98. Also, cable type of hood lock with underdash button controlling the lock mechanism. Price $12.95. Both available from J. C. Whitney & Co.

A complete alarm kit with time-delay switches allowing interval of 10 to 12 seconds before activating the alarm. Includes 6 contactor switches and horn. From Sears Roebuck, at $17.50. A similar system but with siren and key lock switch, price $24.95. From Lafayette Radio & Electronics, a complete siren alarm kit, catalogue no. 11F62015, price $27.85.

Installed prices for car alarms vary considerably according

to type and brand of alarm and locality. A range of from $50 to $80 is about average.

Another siren type alarm is the Ademco Model 109 with improved siren motor designed to avoid burnouts. Complete system for either 6 or 12 volt cars, available from M. D. Kramer Locksmith Supplies, price $24.

A hood lock in which a sprocket chain is gripped by a catch until released by a key lock is available from Kramer Locksmith Supplies, $11.00. Also from the same source is Krooklok, consisting of telescoping tension rods. One hooked end slips over the brake pedal, while the top end hooks over a steering wheel spoke. The lock snaps onto the rod to hold the adjustable sections under tension, preventing movement of either the brake pedal or steering wheel. One advantage of

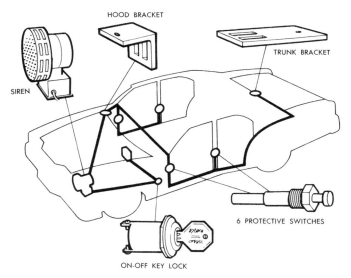

HOOD BRACKET

TRUNK BRACKET

SIREN

6 PROTECTIVE SWITCHES

ON-OFF KEY LOCK

Effective automobile alarm installation covers all doors of the car, the trunk and hood with special plunger type switches. External on-off switch is mounted on cowl, and loud siren is located behind the radiator grille. An interior control switch, either time delay or one that is secretly located, can serve instead of the cowl lock switch that requires use of two keys each time the car is entered.

this device is that it is not necessary to lock the car doors. Price $11.00. A similar lock is available from Sears Roebuck at $9.99. A simple hood lock kit, for certain car models, is listed by Lafayette Radio & Electronics, Catalogue No. 11F53014, price $3.49.

Installing a Siren Alarm. A typical alarm installation involves first mounting a siren or horn under the hood. The space between radiator and front grille usually is suitable, as it would not become an obstruction to servicing of the car motor, and there would be minimum blockage of sound waves. In most cars, a cross member or bar will be found in that space to which the siren or horn can be attached with two screws into drilled holes.

Mount the control switch conveniently near the driver's seat, possibly under the dash where it would be completely concealed but can be located by touch. Many alarm systems use a key switch that is mounted outside the car through a hole drilled into the fender or cowl. The key switch turns the alarm on from outside after leaving the car, and disconnects before the driver enters the car.

Small pin switches are installed on the doors, hood, and trunk; also under the tape player if desired. Other types, such as button, spring leaf or tamper switches, are available to meet any difficult conditions that may be encountered. Installation can be made without damaging any of the car's upholstery.

The electrical connections are made with 16 or 18 gauge stranded wire. Feed two wires through the fire wall to the siren terminals (both these wires are positive live leads and should be protected against grounding). One fused wire connects to one side of the control switch, the other is connected into the car's fuse block. The contact switches are the open circuit type (that is, the contacts are open when the switch buttons are depressed, with the doors closed) and are wired in parallel. One wire of this parallel circuit is connected to the second side of the central switch, the second wire to a good chassis ground. Closing of any one contactor in this circuit will sound the siren if the control switch is "on."

II PROTECTION AGAINST HOME HAZARDS

16

Protect Your Home from Fire

Though you may have taken every rational precaution, a fire can occur almost anywhere in the home as a result of spontaneous combustion, momentary carelessness, defective equipment, electrical short circuit, or other causes. A blaze that has been smoldering undetected for hours can flare up suddenly at 3 A.M., threatening the safety of the sleeping family.

The primary purpose of a fire warning system is to enable your family to escape unharmed in the event of a fire. An alarm system must be absolutely dependable, always on the alert, cover all possible sources of a fire, detect the danger in time, and sound an adequate signal.

Fire warning systems are similar to burglar alarm circuits, the chief difference being the type of detectors used. Where the burglar alarm depends on contactor switches that operate when a door or window is opened, the fire warning system utilizes heat sensors and smoke detectors. There are other ingenious detection devices, but they are for special purposes, such as the infrared flame detector, which is used in locations where fire is apt to develop instantaneously from flammable liquids; a water pressure valve that functions when a fire sprinkler has turned on, and an ionization detector that senses the presence of combustion gases in an air duct.

How System Works. The low-voltage fire alarm circuit is wired always with open circuit detectors, installed in the ceiling of various parts of the house. A panel box contains the central control instruments. The signaling equipment differs somewhat from that of the burglar alarm—instead of a loud outside bell to scare off intruders, the fire alarm uses an indoor siren or horn primarily to awaken the family for quick escape.

Of course, the fire warning can include an outside signal bell and a flasher light, and also be tied into a telephone dialer system that will call the local fire station and give the alarm.

A fire alarm system, by definition, refers to an installation having multiple detectors and a central control box with remote signaling devices. There also are individual self-contained units that are useful for residential protection. One of the most interesting and effective is the Falcon Automatic, consisting of a tank of freon gas linked to a signal horn. An exposed heat sensizer responds to a rated temperature level, releasing the pressurized gas to sound a long blast of the

Low-cost, self-contained fire alarm can be mounted on a wall without any wiring. Unit operated with standard C battery. Stanley Hardware makes this one at about $10.

horn. The Falcon unit, which has the Underwriters' Label, is installed on any wall near the ceiling, needs no wiring or batteries, can function after remaining dormant for years on end without attention, and is regarded as highly dependable. Installation usually can be done in a few minutes with just a screwdriver.

Other self-contained units are the Edwards plug-in home alarm, the Alarmco Watchman, Emerson Electric Company Unit, and others.

As a general rule, it is advisable to purchase only U.L. listed or approved fire alarms and auxiliary components. The alarms described in this chapter are not necessarily all U.L. listed, and some may have certain deficiencies, the chief one being the reliance on small flashlight-type batteries which often are neglected and run down without warning. Also, a number of fire alarms examined were found to be of very shoddy quality, having electrical contacts that are subject to corrosion, fusable elements of doubtful sensitivity, and questionable tension springs. The buyer usually can spot such products if he insists on removing the cover to examine the terminals and contactors. Adhering to a reputable standard brand in the purchase of as vital a product as a fire alarm is one way to avoid serious error. For those who lack the knowledge to make an independent judgment, the best guide is the Underwriters' Label.

Types of Fire Detectors. The basic component of a fire alarm is the fixed temperature heat sensor or detector, sometimes called a thermostat. This is a compact plastic disc, about 2½ inches diameter and barely an inch thick. It is available in two temperature ratings, fused to make contact at 135 degrees for average home locations, and 190 degrees for boiler rooms, in kitchens, near the oven, in the attic, and similar places where occasional high temperature conditions occur normally. The detector is almost always of the open circuit type, even when combined with a closed circuit burglar alarm in the same panel box, which is accomplished by separating the fire alarm circuit to bypass the electromagnetic relay. These sensors do come with closed circuit

Most widely used fire sensor is the fixed-temperature thermostat, such as the Edwards detector shown, fused either for 135 or 190 degrees.

contacts, but only for special installations. Cost of the ordinary open-circuit type is approximately $3.50 each.

This type of detector is rated to protect an area of 20 × 20 feet, or 400 square feet when installed in the center of the area. A single detector normally will serve to cover almost any large room in the house, except perhaps a basement playroom which may be larger and require two of the detectors. There is no limit to the number of detectors that may be installed in any alarm system. For the average home, 8 or 10 detectors will cover most vulnerable locations.

A combination of the fixed temperature with a "rate of rise" detector provides an even more reliable and earlier fire warning. Slightly larger than the fixed temperature unit alone, being about 3¼ inches diameter and having a dome shape, this unit functions when there is a heat increase greater than 15 degrees a minute. The combination unit costs about $7.75 each, and protects up to 2,500 square feet of unobstructed area. Thus this detector is particularly suitable for the basement playroom installation rather than two of the fixed temperature sensors, and at no greater cost. However, it should not be used where there normally may be rapid temperature variation, such as over heat registers or near furnaces.

Locating the Detectors. Fire detectors are installed on the ceiling. The detectors are most effective when located at the center of the room, but that is often very difficult if not impossible. In practice, the location is determined by ease of access for making the wire connections, a desire to avoid damage to the ceiling plaster, and objections to placing the detectors where they will be prominently in the line of vision.

The functional capability of a single thermal detector is

regarded as its "square of protection." This must be under-
stood as the unobstructed or open area of which the detector
is the center. Thus, for a detector rated as providing effective
coverage of 20 by 20 feet, or 400 square feet, this is so only
when the detector is at the center of the area, with 10 feet
of open space in each direction. (See sketch). When the
detector is at one corner of the ceiling, its full effectiveness
is limited to 10 feet from its position, leaving the rest of the
room not properly covered.

When detectors cannot be placed at the center of a room,
they should never be located closer to a wall than 6 inches,
as that part of the room, where wall meets ceiling, is a "dead"
air space which is bypassed by swirls of hot air from a fire.
Where there is no other alternative, a detector may be placed
on a wall, not less than 6 inches or more than 12 inches below
the ceiling line.

How does this work out? In an average-size room, 12 × 16
feet, a detector placed 2 feet from each of the corner walls,
will effectively cover the entire 12-foot area (2 plus 10 feet)
in one side of the room and 12 of the 16 feet of the other side,
providing fairly complete protection. In a larger room, 14 ×
20 feet, while one side is protected, a fairly large area is not

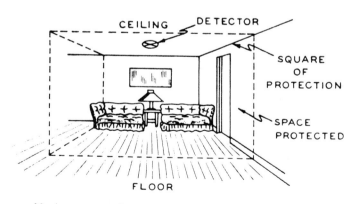

Maximum area of protection is provided by thermal detector when
it is located at center of room, on a smooth ceiling. Open beams or
ceiling joists affect range of effective protection by each detector.

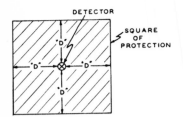

Thermal detectors are rated for "square of protection." This means the distance from the detector to the nearest wall at each side. The letter "D" refers to the space covered by that protector.

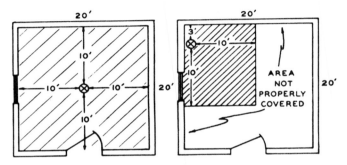

Rated for 20 feet by 20 feet, the detector will cover only 10 feet in each direction, if located at the center of a room. Moved closer to a wall, it will cover only the distance to that wall on one side, and 10 feet to the opposite side, as shown in the diagram.

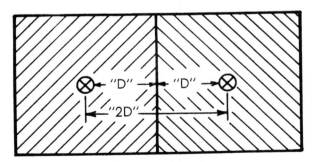

When located in rooms wider than area of protection, figured from the center of its location, a second detector would be required to cover the remaining area of the room.

fully covered, and in that situation it would be advisable to move the detector farther away from the corner, or install a detector at each end of the room. Of course, a single detector in the center of the room is the best answer.

The Smoke Detector. Another important and dependable protective device is the smoke detector. Smoke causes more deaths than fire itself. A slow-burning fire in an upholstered chair may generate enough smoke or combustion gases in a home to cause unconsciousness or death before producing enough flame to activate a heat detector.

Drawing shows construction of smoke detector combined with heat detector. It is mounted on the ceiling.

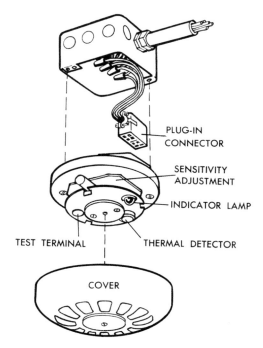

PLUG-IN
CONNECTOR

SENSITIVITY
ADJUSTMENT

INDICATOR LAMP

TEST TERMINAL

THERMAL DETECTOR

COVER

A smoke detector may give early warning before the fire has gained headway. The device operates on a simple principle of optics involving the reflection of a light beam from airborne smoke particles onto a photoconductive cell. A density of 2 to 4 per cent obscuration in the smoke chamber triggers an alarm relay. A smoke detector can monitor an area 60 by 60 feet in size, or 3,600 square feet of floor space, if the device is located at the center position. The average smoke detector is about 6 inches in diameter, projects just two inches below ceiling line, and includes a built in fixed temperature detector to increase the scope of its capability. The smoke detector is easily wired into the same circuit with the heat sensors by a parallel loop in which two wires from the preceding heat sensor are brought to the smoke detector terminals and continued on to the next sensors, then back to the panel box for continuity of the current.

Locating a Smoke Detector. Although the smoke detectors are fairly expensive, a single one of these units may be adequate for basic protection in the average home in which the bedrooms are all on one floor or in one area. The smoke detector is best located in the ceiling of a corridor or other area between the bedrooms and the rest of the house, to

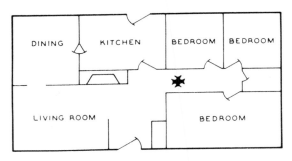

Best location for a smoke detector is the corridor or staircase area between the bedrooms and the rest of the house, to stand guard against the gases from a slow, smoky fire. In homes with the bedrooms all in a single area, just one smoke detector may be adequate.

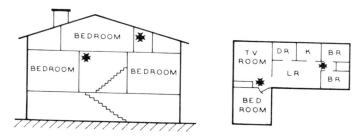

Recommended distribution of smoke detectors for homes with widely separated bedrooms. An important protective detail is that occupants sleep with individual doors shut to provide a barrier against rapid entry of smoke.

provide warning of a smoke condition. This recommendation is based on the assumption that bedroom doors are kept closed while their occupants are sleeping, thus providing a barrier to the smoke that will gain additional seconds for escape from the rooms on the sounding of an alarm.

Additional smoke detectors are desirable in areas which are susceptible to smoky fires, such as the living room, basement playroom, or den. A smoke detector in each bedroom also would be of value to provide additional protection from a slow-burning fire in bedding or upholstery.

Integrated Fire Alarm Panels. One way to be sure of having a dependable fire warning system is to obtain a complete alarm outfit that is manufactured by a well-established and experienced company in the field, and bears the U.L. label. There are any number of fire alarm units available, and it will take careful buying to obtain one that will be dependable. Always bear in mind that you hope the system will never be put to actual use, but you want to be certain that it is always in functioning condition and will not fail should the necessity ever arise.

The control unit is the nerve center of any fire alarm system. Through it passes the electrical pulse that initiates the warning signal.

An engineered, pre-assembled panel box is recommended

because it has all the instruments required. Its complex hookups are fully and properly wired, giving you features that you may not be able to install yourself. A good basic panel box will have a transformer to operate on house current, but with space for a standby battery and provision for automatic switchover in the event of power failure, or an automatic charger to keep the battery in peak condition. Also the panel should have provisions for testing the circuit wires and signaling equipment without sounding the alarm bell; means of supervising the circuit wiring with a trouble signal notifying of any break in the line or connections; and connections to sound both indoor warning horn and an outdoor bell in the event of fire. Pilot lights or dials on the panel door show at all times the condition of the system and the power source.

Advanced functions add to the cost, but you may well consider them worth the investment. One is the telephone dialer, the same type used in the burglar alarm system, with a single dialer coded for sending the correct message for either burglary or fire. Another feature is zoning, which is useful but perhaps not quite essential for a home system. Zoning is intended quickly to locate the source of a fire and is also helpful in localizing any break in the circuit wiring or other trouble. Zoning is discussed further on in this chapter.

Wiring the Circuit. Use two-conductor, solid nonstranded wire, 18 to 22 gauge, for the detector circuit. Twisted wire will be easiest to handle for making the terminal connections. Where some surface installation will be necessary, wire with clear plastic insulation can be used to advantage as it blends well with the background and is barely visible although it does have a glossy appearance.

The circuit starts at the panel box, continues around to all detectors, which are connected in parallel, and returns to the panel box. The panel box is so set up that the circuit wiring forms a complete loop. The loop back to the panel permits testing the condition of the wires without sounding the alarm. When wiring the circuit without a panel box, start with one wire from the battery and another connected to the

Edwards fire alarm unit contains its own transformer. Circuit wires are low voltage, like bell wire.

Wire from alarm may be simply tucked along the top of base molding with insulated staples or plastic tacks. Alarm is powered by plugging into regular house current.

constant ringing drop (but do not connect the battery lead until the circuit wiring has been completed and tested). There is no need for maintaining polarity. The other battery terminal and a lead from the constant ringing drop are connected to the signal horn. (A second wire from the battery also goes to the "drop" terminal of the C.R.D. as shown in the diagram for the burglar alarm system.) This is the basic circuit.

A testing setup is interposed in this circuit arrangement. A selector switch is installed on the line connected to the C.R.D., so that when the test button is pressed, the C.R.D. is disconnected. Instead, a buzzer is sounded. The two ends of the circuit loop permit this. One wire is soldered to the test terminal of the selector switch, the other wire to a buzzer, which is also wired to the battery. Thus the buzzer test will show the condition of the circuit wires—bypassing the open detectors, without sounding the horn through the constant ringing drop.

When wiring the detectors, it may not be practical to keep them all on one wire loop because of distance and location. The use of two or more loops, all connected to the proper terminals at the panel box, serves the same purpose as having just one wire circuit.

Installing the Alarm Devices. The detectors are installed in a ceiling, preferably at the center of the room or at the head

Almost all thermal detectors are of open circuit type, thus are wired in parallel. Some have double sets of terminals, as shown, others have single terminal screw on each side. The detectors may be wired into a loop from the alarm circuit, or the wires continued on to the next detector—in either case, fusing of the detector closes the circuit and triggers the alarm.

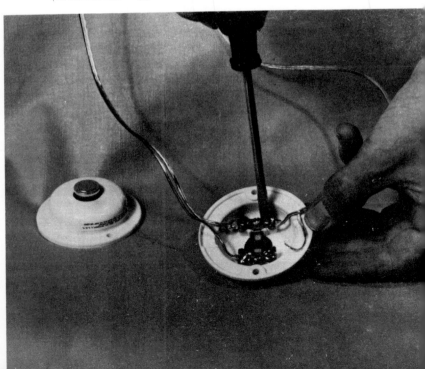

of the staircase. In some homes, particularly the one-story ranch type, the ceilings of most rooms are easily reached through the attic. A tiny hole is drilled using the 12 or 18 inch long bit, to pass the two-conductor wire leads down through the ceiling to be connected to the detector terminals. The detector then is attached to the ceiling with two long, thin screws. In most attics the main obstacle will be getting through the insulation batts packed into the floor joists, and every effort should be made to avoid disturbing the batts. Instead, the hole is drilled upward through the ceiling (first using a carbide bit to penetrate the plaster) but measure the distance to avoid drilling through the joists. When the bit has moved through the insulation, it is removed from the drill chuck and a thin steel wire taped to the end of the bit. When the bit is drawn down through the packed fiber insulation, the wire that is attached is the means for pulling the alarm circuit leads through the ceiling. The low-voltage conductors are thin and will pass easily through a $1/4$-inch hole. The wires are connected to the two terminals of the detector.

Detector installation in lower-floor rooms of two-story homes is more challenging, but not impossible. The best time to install the fire detectors is just before the rooms are to be redecorated; any holes that must be drilled or cut into the ceiling then can be patched easily, and the painting will completely cover up the repairs. But even without that fortunate timing, many of the detectors can be installed without causing any damage.

Adjacent to almost every room there is a closet inside which some plaster may be broken near the ceiling corner. Then a $3/4$-inch hole is drilled in the joist nearby. A small hole is drilled in the room ceiling at the corner within a foot or so of the closet. A wire tape (snake) is inserted from the closet through the joist and towards the ceiling hole. A stiff wire (part of a wire clothes hanger) with the end bent into a hook is put up through the ceiling hole in an attempt to snag the wire tape from the closet.

This method works quite often, but not always, so don't be too disappointed if you can't make the contact. One alternative to this method is to chop away some plaster at the ceiling corner of the room, drill through the joist obstruction, run the snake from the ceiling location hole to the cove opening, then through or under the joist and into the closet. Patching the small plaster damage is quite simple. If there is a cove molding at the ceiling line, that makes things more difficult, as removal of the entire molding will cause quite a bit of damage. But tiny holes drilled in the molding are easily patched with plaster so they are invisible.

Where hot air ducts run at ceiling level, the circuit wiring can be snaked through, inside the duct if necessary. Basement and attic locations of the detectors are usually without complications, the wires can be drawn from floor to floor through holes in the closets, alongside plumbing lines, heat-ducts, or through spaces in the exterior walls.

Be sure to include a detector in the kitchen, furnace room, playroom, living room, basement workshop, attached garage, utility room and storage pantries. In large and remote rooms, such as a basement playroom, use the combination rate-of-rise and fixed temperature detector, and if you can, a smoke detector.

Zoning the Detectors. A zoned system is different only in that the detectors are bunched into groups and wired in separate loops, each connected in the panel box to an individual pair of terminals so that when there is a trouble signal or any thermostat initiates the alarm, the panel box will show the particular area involved. In a typical installation, thermostats in the basement may be connected to Zone 1, those on the first floor to Zone 2, those in the attic to Zone 3, those in the attached garage and room extension to Zone 4. While zoning may not in itself be an important feature, practical problems of wiring the circuit may require separate zones since it usually is not possible to run one continuous circuit to all parts of the house.

Power Source. Fire alarm systems are sometimes internally powered units, depending on a 6-volt lantern type battery to sound the alarm. This battery is rated for a shelf life of one year, although it may begin to fail within six

Installing Detector

Installation of fire detector is facilitated in basement by removal of the ceiling cove molding. Hole is drilled first into the ceiling tile and wire tape snaked through and snagged at the wall opening by a second steel wire. The alarm wire then can be pulled through by withdrawing the steel tape.

The detector is attached to the ceiling with two screws. Its size, shape, and color make the unit almost unnoticeable on the ceiling.

months. A better arrangement, which is available with some alarm panels, uses a 6 or 12-volt transformer off the house current, with a standby 6-volt battery for automatic switch-over if the house power fails. An effective variation of this arrangement is a battery charger that automatically keeps the battery at full power.

Authoritative opinion favors transformer current in preference to reliance on a simple battery setup. The view is that power failure is infrequent and usually for a short period, occurs mostly at times when the resident would be aware of that condition, and there is only minimal possibility of power failure coinciding with an actual fire occurrence. In contrast, battery failure is common, as there is a tendency to neglect

Fire alarm system by Alarmco has four circuit zones for quick location of alarm signal source. Test switch causes bulb to light instead of ringing the alarm horn.

charging or replacement, so that the battery could be non-functional at the time it is called upon to sound the alarm. The combined transformer power-plus standby battery, or the constantly charged battery, meets the power requirements satisfactorily. Of value also is a meter that shows at all times the condition of the battery. The meter itself costs barely $10, and is worth the extra investment.

The transformer used for alarm purposes should be rated for not more than 30 volts. This transformer is usually installed permanently into the electric service box, but it is permissible to plug the prime current cords of the transformer into a receptacle — but be sure it is not one that can be disconnected by a wall switch. The transformer plug should be secured in such a manner that it cannot be inadvertently withdrawn from the receptacle.

Testing the Trouble Indicator. Periodic testing is essential to maintain dependable function of the alarm system. Fire alarm installations without the control panel also should include some means of testing the system without sounding a false alarm. Wiring for the test buttons is described previously in this chapter. The National Fire Protection Association recommends that home fire alarms be tested weekly, and that a definite day each week be set for this routine.

All alarm panels have some form of test switch. In simple, self-contained systems, a button merely checks the operation of the alarm signal. The more advanced panels have a selector switch which permits silencing the main signal bell while the detector wiring and panel connections are checked by sounding a buzzer from the standby battery. Pilot lights on the cabinet box indicate that the system is "off" so that there is no chance of forgetting to restore the alarm. The buzzer test signal is ideal also for holding fire alarm drills.

A supervised system is one that has a trouble signal to warn of a break in the detector wiring or other fault. It also sounds a warning when the house power fails and the system has switched over to standby battery.

Signaling Equipment. A distinctive sound of an indoor signaling device is desirable for the fire alarm, so that there is no mistaking the nature of the emergency. For this reason, fire alarms should have signaling equipment separate from those of the burglar alarm. Loud, earpiercing horns or sirens have been found suitable for this purpose. But it is essential that the sound be entirely different from auto horns and other traffic noises to which we have become accustomed. An indoor horn and outdoor siren, to notify neighbors or attract the attention of police, make a practical combination, but be certain that your power supply is adequate to sound both of these devices simultaneously. Other types of signal devices are single-stroke bells, chimes, vibrating bells and sirens.

The recommendation of the National Fire Protection Association is that all alarm sounding devices for home installation shall be rated not less than 85 decibels at 10 feet, and that the alarm signaling devices shall be so located as to be clearly audible in all bedrooms with all intervening doors closed.

Typical of the available signaling equipment are the following:

Alarm Devices Manufacturing (Ademco)
A-518: dual indoor warning horn, 6 volts in plastic case, $7^{1}/_2 \times 4^{1}/_2 \times 3^{1}/_2$ inches.
A 65-6 and A-66, Box A-68: siren, 6 volts DC, or from AC transformer, or 110 volts AC.

Indoor A-UF8-4 and UF8: indoor bells, 4 to 6 and 6 to 12 volts.

A-WF8-4 and A-WF8 Outdoor Box: outdoor 4 to 6 and 6 to 12 volts, A 211.

Edwards

6100-4/8: 4-, 6-, and 8-inch DC vibrating bell, 75 to 82 decibels at 10 feet, $11.50 to $19.

4-375-L: horn, 6 to 12 volts, 95 decibels at 10 feet, approximately $35.

33-4: single-stroke bell, 6 volts, 76 decibels at 10 feet, approximately $23.

34-A: buzzer, 24 volts, 70 decibels, approximately $16.

Rittenhouse

RS 644: alarm horn.

RS 624: outside warning light and horn.

RS-625: indoor alarm horn.

Lee Electric

No. 1202: outside 8-inch underdome bell in metal case, $40.00.

No. 1204: 6- and 12-volt siren, in metal case for outdoor installation (requires car-type storage battery), $44.50.

3200: vibrating bell.

4500: horn, 24 volts, $16.10.

4400: siren.

Lafayette Radio and Electronics

12F 74067: bell rated 80 decibels, $8.75.

AUTOMATIC SPRINKLERS

Automatic sprinkler systems are required by all fire codes and building ordinances for industrial plants, hospitals, offices, major stores, and other public buildings. However, sprinklers have never become established as a fire prevention and protective installation in homes. There are some pros and cons on this subject, some of which will be examined here.

Sprinklers are connected into the water line and usually are installed in or just below the ceiling. The sprinkler "head" has some type of fusible link, usually consisting of a special solder that melts at a predetermined temperature, generally 160 degrees for areas where normal temperature does not exceed 100 degrees. When heat from a fire melts the solder, a valve opens and the water sprays out. A metal shield deflects the spray into a specific pattern that will drench the area to extinguish or contain the fire. The sprinkler can function in conjunction with a water flow detector to sound a fire alarm.

Sprinklers vary considerably in design to meet specific problems and installation conditions, including a compact and attractive unit that is installed flush with the ceiling, so that only the fusible link projects one inch below the ceiling level.

A standard sprinkler head, operating under minimum discharge pressure, will spray a circular area about 20 feet in diameter. This area is about equal to that of a good-sized room, and probably one-fourth to one-half the total size of a home basement playroom. Thus one sprinkler head would be adequate for any room in the house, while two would be required to fully protect a playroom. One sprinkler each might be needed in such locations as the kitchen, den, workshop and storage closet. It is doubtful that a water sprinkler is a suitable installation for a furnace room where the fuel is gas or oil.

Two types of fire sprinklers are activated differently. One has a soldered fuse that melts, the other has a bulb of liquid that expands with heat.

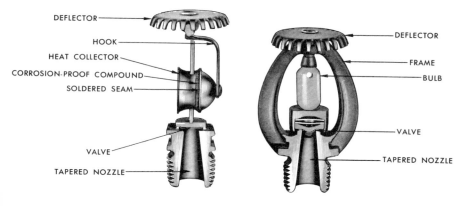

DEFLECTOR
HOOK
HEAT COLLECTOR
CORROSION-PROOF COMPOUND
SOLDERED SEAM
VALVE
TAPERED NOZZLE

DEFLECTOR
FRAME
BULB
VALVE
TAPERED NOZZLE

The Pros and Cons. Objections that are cited regarding residential sprinkler installations are that most homes do not have sufficient water supply and pressure to douse a raging fire; that the resulting water flow if a sprinkler were activated would cause extensive damage, often exceeding the damage caused by the fire itself. (Insurance policies that cover the fire damage may refuse claims for water damage caused by the sprinkler.) And another point is that a sprinkler installation in the home would involve considerable cost.

Taking the last objection first, the answer is that only a limited sprinkler installation is envisaged for the home, providing protection only in particularly vulnerable areas such as the basement playroom, workshop and storage rooms, all as a supplement to a complete fire alarm system. Cost of such an installation would be quite small, possibly under $100, if you do the installation yourself. The sprinkler heads, purchased at plumbing stores cost only about $4 each, and just three or four would be needed. The only other costs would be for the pipe or copper tubing, the quantity depending on the lengths of the runs, a couple of shut off valves, and the fittings.

An authority on fire prevention, Mr. Norman J. Thompson, author of the book *Fire Behavior and Sprinklers,* has stated that there are many situations where installations of sprinklers in private homes would be advisable. He agrees that the water pressure in private homes is too limited to supply water for more than a very small number of sprinklers; however, it is a good assumption that only one or at most two sprinklers would be called into function at any time, since there would be no more than that number in the initial fire area and thus the entire water supply would be concentrated at that spot.

The point made here is that the sprinklers are used in conjunction with a fire warning system, that they are intended primarily to slow the development and spread of a fire while the family has been alerted to the danger and has had time to escape. As to water damage, that is a personal and individual question—some basement areas have little that can

Sprinkler head may be placed into tee fitting on any waterline, either pipe or tubing. Largest size water line in house should be chosen. Tee is cut into line with a union coupling if necessary. Be sure to place the sprinkler in the correct position. Some are designed as pendant (extending downward), others are upright and the spray deflector is above the head.

be damaged by water; others are crammed with "valuables" and perhaps the stress here should be on emptying the basement rather than worrying about how much loss would be occasioned by the water used to extinguish a life-threatening fire in the home.

Installing the Sprinklers. Strict ordinances cover fire sprinkler installations in commercial and public buildings, specifying the source of water, pressure required, number and location of pumps, size of pipes and location of sprinklers, valves, hydrants and other technical details.

A home sprinkler installation does not have to meet any of these requirements. However, some localities require a permit for any plumbing work involving the water lines. Installation involves tapping off a branch line with a tee fitting, and installing the sprinkler head into the line at the desired location. It is very important, if you are to obtain the protection you seek, that you use sprinklers of the right temperature rating, and carefully follow the manufacturers' instructions on handling and installing the sprinkler heads. The sprinklers are all clearly marked as to the position in which they are to be placed. Those reading "upright" are installed with the spray deflector above toward the ceiling; those reading "pendant" are installed with the deflector downward toward the floor. Some types can be installed in either position, while another is marked for "sidewall" installation. Sprinklers must be handled very carefully to avoid damage. (The best way is to keep them in their original containers until ready for installation.) Any sprinkler that has been dropped or otherwise subjected to possible damage should be returned to the factory for examination.

Another word of caution: be careful when moving large objects, such as ladders, that might bump a sprinkler and start a torrent. And always remember that while the sprinkler can stand guard alertly without further attention for many years, it is always possible for a sprinkler head to suddenly open up without apparent cause. That is one of the chances you take for the greater safety that the sprinkler can offer. Is it worth it? That's for you to decide.

There are no comprehensive records regarding the experiences of homeowners who have installed the system. A survey of various sprinkler manufacturers regarding the home installation situation has brought almost identical responses: while every commercial building is protected with a sprinkler system, homeowners have not been convinced that the expense of such an installation is warranted. Many firms have discontinued efforts to interest the homeowner, and so far as is known, only a limited number of such installations have been made. However, the National Automatic Sprinkler and Fire Control Association is studying the subject anew, and there is reason to believe that some day this type of fire protection will come to be regarded as "worth the expense" because it *can* save lives.

CAUSES AND PREVENTION OF FIRES

Aside from alarm systems, home fire safety calls for two things: first, thoughtful and thorough steps to remove existing and potential fire hazards; and second, a clear plan of action, ready to meet the various fire emergencies that could arise in your house.

Fires are said to be caused mostly by carelessness. That may seem too broad a statement, but it at least can encourage us to see how we can prevent fires.

The United States, with almost 2,000 residential fires every day, has the worst fire loss record of any major country for which accurate statistics are available. Britain, with approximately one fourth of our population, had fewer than ten per cent as many fires, and an even smaller proportion of fire-related fatalities. Fire prevention is taken seriously in Europe, and the strict regulations evidently have been effective.

Some reminders are listed here of common household conditions or personal habits that require constant correction to maintain fire prevention.

Conglomerate Pileups. Avoid accumulating flammable materials such as newspapers, magazines, old mattresses, broken furniture, etc. Most of it will never be of any use to you. Start an immediate cleanup, discarding everything that is not obviously needed, or does not have a definite and safe storage place. Certain charitable organizations, like the Salvation Army or the Lighthouse, will send a truck to pick up some kinds of objects.

Keep books tightly stacked in their bookcases. Bills, letters and other papers must go in a steel drawer or filing cabinet, never allowed to clutter desks or counters.

In the Workshop. Good workshop practice includes frequent cleanup of wood scraps and sawdust—using a vacuum cleaner for the latter behind panels or boxes, where the highly flammable dust accumulates. Discard oily rags outdoors immediately, make sure all paint cans are tightly sealed, and keep on hand only small quantities of paint thinner, rubber cement, and similar flammable liquids, all in tightly sealed metal cans or polyethylene jugs such as the kind in which chlorine bleach is sold. Keep a dry chemical extinguisher (suitable for all types of fires) handy in your workshop. A pail of sand or water also would be a useful firefighting aid.

Dangers of Smoking. Much as smoking has been condemned as a health hazard, it is an even more direct threat to life from fire. More than one of four fire fatalities results from careless smoking, according to a study by the Metropolitan Life Insurance Company. Smoking in bed was a leading cause of fires and fire deaths, and this practice should be ruled out in every home. Additionally, fires were caused by the emptying of ashtrays with still smouldering cigarettes into a waste basket. Provide ashtrays of adequate size, not those dainty porcelain things which can barely hold a cigarette in place, and are always overflowing. As a regular routine, empty ashtrays into a "metal butler," a tin box with a hinged lid, which is allowed to stand for hours before being emptied into the trashbin. If a live cigarette has fallen behind uphol-stery cushions, don't relax until it is found, or assume that the cigarette has been extinguished; remove the cushions and dampen the area, then remain alert for at least six hours to the possibility of treacherous smouldering.

Electrical Safety. Keep the fuse box clean and closed, re-place blown fuses only with those of same amperage (15 amps. for a lighting circuit, 20 amps. for an appliance circuit, 30 amps for heavy duty three-wire circuit serving a range, air conditioner, heater, etc.). The wire gauge determines the current-carrying capacity of a circuit, and just changing a fuse to a higher number will not increase the electrical ser-vice; instead it will invalidate the safety-valve function of the fuse so that an overload or short, instead of blowing the fuse, will overheat the wires, burn off the wire insulation, and cause dangerous sparks. Stringing extension cords under rugs, tying them on nails, or using excessively long cords of insufficient gauge also constitute fire hazards.

Heating and Cooking. Have your entire heating system inspected and adjusted by a service mechanic before each winter season. Keep portable heaters away from curtains, bedding, and other combustibles, and be sure that all flame heaters, such as gas, are vented to the outside. Make annual checks of the chimney and flue pipes to be sure that masonry separations are intact and that there is no contact with the house framing members. Kitchen exhaust fans collect grease that can flare into flame. Regular cleaning of the fan motor, filters, and duct is important.

Attic Fan Draft. An attic fan can become a lethal instrument should a fire start anywhere in the house while the fan is in operation. The air currents will speed combustion, turning a small fire into a holocaust in seconds. Fan louver shutters can be equipped with fusible links, and the fan power circuit fitted with an automatic cutoff switch actuated by a fire detection system, to eliminate these hazards. These safeguards are required installations in some municipalities.

Close the Door. All bedroom doors should be kept closed at night; this limits the draft and thus slows the growth of any fire. A closed door also holds back toxic gases that are even more deadly than the flames. The door leading from the basement to the upper floor should be tight fitting and always kept closed at night. But make sure, by actual test, that the fire alarm signal can be heard through the closed doors. If not, install a louder horn or siren, or individual signals for each room.

Rules on Matches. Matches are so commonly used that we are no longer wary about them. Safety book matches are preferable to the old style large wood type, but always close the cover before striking. This can save you from a very painful burn and prevent a fire which would start if the open matches flared up and, by reflex action, were dropped onto flammable material. When searching in a dark closet, or when seeking the source of a gas odor, always use a flashlight, never lighted matches. One of the prime rules about matches is to keep them out of the reach of children. This may be difficult to do in the busy home kitchen, but very necessary. Store matches always in closed, non-flammable containers, metal boxes or large glass jars, and never close to the kitchen stove.

Christmas Tree Fires. Lighted candles have no place on Christmas trees. Evergreen needles are highly flammable, and even the smallest start of combustion can become a flash fire. The question is often asked why it was possible to use candles in the days before electric lights. In those times, most homes were on farms or not far from wooded areas. The trees were freshly cut, less likely to burn. Even so, Christmas tree fires were much more common then than now.

With electric lighting, there still is a possibility that defective wiring will ignite dry, flammable branches. Spraying the tree with ammonium sulfate or other fire-resistant chemical is a worthwhile precaution. Use only U.L.-labeled tree lights, plugged into a receptacle that is on a 15-amp. fused circuit. Never plug into an appliance circuit, such as one used for the air conditioner, which has a higher-temperature fuse, and avoid using exceptionally long extension cords. Tree lights should be disconnected before you leave the house or retire for the night—a timer clock that switches off the lights automatically is a satisfactory means for accomplishing this. When you buy the tree, make sure that it is not more than a week old, the needles still supple and moist, and no sap has run at the stump. Discard the tree no later than New Year's, and sooner at any signs of brittle dryness. Remove gift wrappers promptly, and discourage smoking near the tree.

PLANNING THE EMERGENCY EXIT

Look around and try to visualize what would happen if, despite careful prevention, a fire were to break out in your home. Get this message across to every member of the family: if there's a fire, get outside, at once! Don't stop to fight the fire, or call in an alarm, or waste precious moments trying to salvage any belongings. The alarm can be sent from outside by a neighbor's telephone, or from a nearby alarm box—and you should know where it is. Do you?

The means for exit from every room, and every part of the house, should be established and proved by actual test. Routine fire drills are conducted at schools, aboard ships, in many factories and public buildings. Shouldn't they also be held at home, at least once to test the entire escape plan, making sure that no detail has been overlooked that can create a blockade in an emergency?

How would anyone get down from an attic room if the

stairs were aflame? Are your storm windows easily opened, or would they become a barrier to a youngster caught in an upstairs room? Do all doors unlock easily and open without binding? Is a window air conditioner obstructing an only exit? Are packages ever left on stairways? If you live in an apartment house, have you located the emergency stairs? Is the fire escape kept clear all the way down? Does the building have fire alarm bells?

Home Fire Drills. Figure out two ways to reach the ground from each bedroom, the basement, and recreation rooms. Make floor plan sketches of your house, and check off each area as the plan is worked out. Mark arrows showing two escape routes from each room.

With the family assembled, start the fire drill at any room. At a signal, all members leave by the nearest way, going directly outside. Then they return and try the alternate way, out a window if necessary. Everyone follows suit, the women and children and old folks included. Do this complete test at least once—you'll soon see whatever is missing, such as a handy rope ladder upstairs, or any faults in the plan that must be corrected.

If parts of this exercise seem silly, or even dangerous, just think what it might be under the frenetic conditions of a real fire. The test, on the contrary, should be quite safe and even lots of fun for everybody, when conducted under calm conditions and with helpful assistance available for anyone running into a difficulty. Go through with it, by all means, and you may discover certain conditions that were not apparent at first thought, and that need attention.

Two Escape Routes. The regular exit from a room can be blocked by heat, smoke or flames. An alternate way out of every room is necessary—through a window in most cases. Some windows lead onto a garage roof, or a porch deck, providing a safe place to wait for help or an easy way down. If such an exit is lacking, a rope ladder is essential. Windows must be operable, including their storm sash, not only by the man of the house, but by children also. They must not be blocked by encumbrances such as air conditioners, or by permanently fastened metal guards. Large planter boxes are permissible if they are so placed that they can be pushed out of the way, or shoved overboard.

Basements may present problems if there is but one door, and the windows are too small or are barred. If no other way is available, it may be worthwhile to install a separate hatch-cover outside door, or enlarge one of the windows with a deeper well. In place of fixed bars, a metal gate with a simple catch, or a grating on the window well that is hinged and can be lifted from below, may be the answer.

WHAT TO DO IN CASE OF FIRE

When an alarm bell, or the smell of smoke, lets you know about a fire in the house, correct action will help slow the spread of the fire and assist your escape. Some valuable suggestions on what to do when there's a fire are offered by the National Safety Council.

Door Check. In a fire, never dash out of a closed room without first testing the door. Feel the door near the top with the palm of your hand. If it is suspiciously warm, keep the door closed, as superheated air and gases can rush in and overcome you in seconds. Even if the door feels normal, don't throw it open. Rather, brace one foot and hip against the door, while holding a hand above your head across the door and the jamb, then with the other hand ease the door open very slightly. If there's a hot draft or pressure against the door, slam it shut at once. You will then have to wait in the room or resort to an alternate means of leaving.

Out the Window. Leave the room door closed before opening a window, otherwise the draft can fan the fire. If the window provides access to a garage deck or porch roof, you can get out and work your way to the ground, or remain there until help arrives. If you can't open the window or remove the screen, pull out a dresser drawer and use it to smash the glass or screen, then use the heel of a shoe to clean away any remaining glass shards before going through.

In two-story homes, the second-story window may be fif-
teen or more feet above ground. A rope ladder in the room will
permit quick exit, even for elderly persons, and prevent
broken bones. But it still may be necessary to go out the
window even if there is no ladder. But drop, don't jump!
Ease yourself backwards over the window ledge, until you're
hanging down while holding onto the sill—thus your feet are
much closer to the ground, leaving a drop of perhaps only
10 feet. If there are bushes below, aim for them by using
your knees to push away from the wall as you let go. Bend
your knees and roll to a side as you land, to break the impact
of the fall.

Much as elderly persons may be timid about using an
unsteady rope ladder, the needs of the moment will over-
ride their caution. But practice with using the ladder, even
just once during the planning test, will make all the difference. Wheelchair invalids present a more difficult problem, and the answer may be found only by studying the various possibilities that can be utilized, including a sling-type
carrier similar to a ship's buoy chair. A strong hook attached
to a nearby tree or post may make this emergency escape
possible.

Heavy Smoke. When trying to move through an area that
is filled with smoke, crouch low as you hurry along, or crawl
if necessary, since the smoke will be less dense closest to
the floor. Holding a wet towel or pillow over the nose and
mouth will be helpful. Move along the walls of a room, which
will guide you to a doorway; avoid the center of a room, where
the floor is more subject to collapse.

Meeting Place. All members of the household should gather
at a prearranged meeting place, which you should select now,
as soon as they get outside, making sure that everyone is
accounted for. Don't wander around. Anyone leaving the spot
to sound an alarm or for whatever reason should first notify
the group.

Reporting the Fire. Every member of the household should
know where the nearest alarm box is located, and also the

possible places to telephone if forced to flee the house. If the alarm is pulled at the box, wait there until the fire equipment arrives, or if you must leave, post someone else at the box to direct the fire engines.

FIREPROOFING

Many fire hazards can be eliminated or reduced by fireproofing. Use non-flammable materials wherever possible, such as ceramic tiles, asbestos board, concrete. Some materials are fire resistant, which means they are rated to resist burning for a specified length of time. Chemical treatment reduces flammability of such items as draperies and upholstery materials, even wood.

Asbestos board is an excellent fireretarding material, used for partitioning a furnace room or similar fire-susceptible facility. Other places for asbestos board partitions are the basement office, laundry room, and security closet. The asbestos board is made of portland cement and asbestos fibers, inorganic materials that are rotproof and noncombustible. The panels come in ⅜″ and ½″ thicknesses, and are generally available in 4 x 8 and 4 x 10-foot panels. Asbestos board is extremely heavy and inflexible, and must be handled with care to avoid cracking or chipping the corners. Any damage, however, may be corrected by filling the wall space with a mortar of cement and asbestos powder.

Gypsum wall panels are rated fire resistant for specific periods, depending mostly on the thickness of the panels. Hardboard and some coated plywoods also have fire-resistant ratings. Gypsum partition block and concrete blocks make more solid, soundproof and fire-resistant partitions.

Roof Coatings. During the recent large coastal forest fires, many homes with wood shingle roofs caught fire when windborne embers landed on the roofs, often miles away. One protective measure has been to resurface the roof with noncombustible asbestos-cement shingles, sometimes known as mineral wool panels, applied directly over the wood shingles. Another widely used material is the Barrett Bar-Fire roofing shingle, an extra heavyweight asphalt shingle with coarse mineral granule surfacing backed by layers of asbestos fibers, vermiculite and mineral granules in asphalt binders. The Bar-Fire shingles have been designed for fire safety and bear the Underwriters' Laboratories Class A rating for roofing material.

A little fruit jelly spread on the roof when the fire season rolls around will help protect houses from sparks and flames, a forest products specialist of the University of California Agricultural Extension has suggested. The fruit pectin jelly, stated to cost about 15 cents a gallon, is brushed over the shingles. Mixed with water, two gallons of the pectin will cover 100 square feet of roof. Although the coating washes off with the first rain, this is said to be no problem since the fire period coincides with the dry summer season.

A fire-resistant variety of wood shingles is now available, the agricultural specialist reports, which has "built-in" chemical treatment that does not wash out in the rain.

Forest Fire Hazard. Homes in wooded areas that are imperiled almost annually by major forest fires can improve their fire security by corrective landscaping. This requires, first of all, an open space around the house. State law in California stipulates complete brush clearance within 30 feet of the structure, while local fire codes in areas of extra hazardous conditions specify partial clearance up to an additional 70 feet, allowing only a limited number of low specimen shrubs to remain. Strict regulations apply also to ground cover plantings around the buildings.

There is no such thing as a plant that will not burn, the California Agricultural Extension Service points out. Recent major fires, however, have shown that plants on well-watered landscapes do not burn as readily as dry plantings. Tall trees often prevented flying embers from reaching buildings, while irrigated ground covers, like ivy and iceplant, do not readily carry fire if properly maintained to prevent accumulation of litter.

Plants recommended for landscaping that require little water and do not normally create fire hazards are the yuccas and cacti, which retain water during periods of drought. It is said that one or two irrigations in midsummer may make the difference between an extremely flammable plant and one which will not burn readily.

Recommendations for fireproofing a forest home, issued by the Agricultural Extension Service of the University of California, include:

- Remove small trees and brush, leaving only widely spaced, larger trees.
- Prune the lower branches of remaining shade trees, clearing the trunk to a height of at least 15 feet, to prevent ground fires from spreading to the tree tops.
- Debris from the trees, brush and branches are to be burned in small piles only during the wet season. Obtain a burning permit where necessary.
- Clear a fire trail, 6 feet wide, near the outer edge of the fireproofed area, to slow down a running ground fire.

FIGHTING THE FIRE

A controversy has been raging for a long time over fire extinguishers (that is, whether to try to fight a residential fire, or just run and wait for the professional firefighter). Most experts, including the Engineering and Safety Department of the American Insurance Association, urge that firefighting be left to the professionals. They point out that there are many types of extinguishers and that using the wrong one on a particular fire could be disastrous; that the extinguishers purchased by the average layman are generally inadequate in capacity and so tricky that they cannot be used properly under difficult conditions; that some extinguishers generate deadly fumes; that "small" fires very quickly become big blazes; that time lost fighting a fire should be used to turn in an alarm and to leave the premises, and that often a fire that was thought to be quenched flared up again after many hours.

Other interested observers express the opinion that a third to half of all fires can be snuffed out quickly, before firefighters could arrive and much damage is done, if the correct type of equipment is available at handy locations around the house. They point out that many local codes require that a pail of sand be kept in the furnace room, buckets of water in some building areas, and that Coast Guard regulations require an extinguisher be kept on boats. More training of individuals, it is argued, would help cut down the extent of fire loss.

The generally accepted rule is that if the fire is small, caught at the very beginning, efforts may be made to fight it — but only after an alarm has been sent in, all other persons have left the house, and you have a protected escape route open. Be sure you know what extinguisher to use, where it is located, and how to use it. If the fire is stubborn, don't continue. Get out fast and let the professionals handle it. In the garage, a car, or on a boat, firefighting takes on a different aspect that should not determine the attitude towards home fires.

The right type of extinguisher must be at hand (and you're not going to be able to look it up in a book or even read the instructions on the label). It must also be adequate to the task, and in operable condition. All of which indicates that too much dependence should not be placed upon an extinguisher doing the job.

After years of disuse, a forgotten and neglected extinguisher may have lost pressure to the extent that it cannot function when needed; foam extinguishers must be kept from freezing, aerosol cans can lose their gas pressure, small "fire bombs" have extremely limited capacity and effectiveness.

Not just any fire extinguisher will do in a particular place. Make sure the type fits the kinds of fires it may be expected to be used on, as with the CO_2 unit here near the furnace.

Know Your Extinguishers. The information is on the label so learn it well. Locate each type in places where the type of fire may arise for which it is suitable.

CLASS	TYPE OF FIRE	TYPE OF EXTINGUISHER
A	Ordinary combustibles	Water, dry chemical, foam
B	Greases, paint solvents, thinners	Dry chemical, carbon dioxide, foam
C	Live electrical	Dry chemical, carbon dioxide

The multipurpose dry chemical (monoammonium phosphate) extinguisher is excellent for all types of fire. Gas pressure in this type extinguisher spurts a powder that effectively blankets the fire, but leaves a messy residue. Recommended sizes for homes range from 2½ to 10 pounds. If you don't have one of these handy in an emergency, use common baking soda, found in almost every kitchen, which is sodium bicarbonate. Another handy home extinguisher is a simple garden hose, permanently attached to a faucet and conveniently coiled on a wall bracket to prevent curling. The hose may be used only on Class A fires, which are fueled by wood, cloth, paper, and other ordinary combustibles. Never use water on grease or electrical fires, which are Class B and C.

Don't Blow Up. Many of the ordinary items that are used around the home have a tremendous explosive potential. While safety provisions have been built into many of these products, the danger is that long use leads to the kind of excess confidence that relaxes caution. Some dangers are:

Pressure Cookers. When they first came into wide household use, there was quite a series of "incidents," sometimes only causing a frightening shock and food-splattered ceiling, but there were also some serious injuries. Subsequently, many housewives abandoned the speedy cookers, while others became more adept at handling them. Actually, the cookers do not develop any great pressure, and there's no doubt that they are useful in the home. But they must be handled correctly. Instructions for safe use are simple enough, really: make sure that the steam vents are clear and functioning, and do not use the cooker for thick foods that will surely clog the vents, such as pea soup. Hint: keep an ear alerted for the continuing steam exhaust sound—if it stops, shut off the heat quickly.

Charcoal Lighter Fluid. Impatience to get the barbeque coals going can prompt the householder to squirt large quantities of lighter fluid over hot coals. The fluid is a form of benzene, and on a number of occasions the can has exploded. Recommendation: an electric starter will get the barbeque coals going quicker than the chemical method, and more safely. But if you do use fluid, never use it on charcoal after a first attempt has been made to light the fire. Use it only once, and move the can away before lighting the fire.

Lawnmower Fuel. A more common explosive hazard comes from refilling the fuel tank of a motorized lawnmower while the engine is hot. If you run out of gas, always wait until the motor has cooled, then refill the tank out in the open while using a funnel to prevent splashing the fluid over the engine. Keep a lookout to be sure there is no one nearby smoking. Store mower fuel in an approved gasoline safety can, tightly capped. Do not bring the gasoline can into the house.

Oven Safety. Modern ovens have a safety switch that prevents flow of gas if the pilot light is out. But don't depend on it entirely—after turning on the gas, wait at the oven until you're sure that the burner is on; otherwise close the gas knob again. Ovens that do not have an automatic pilot should be lighted with a match at once. Allowing the gas to flow while you take care of some chore may bring a blast that can wreck the kitchen.

High-Pressure Gas. The main gas line entering your home may have high pressure flow up to the meter. Take every precaution not to bump into this pipe when moving heavy furniture, and never hang articles from this pipe. Build a protective cage around the pipe, keeping access to the valve always clear. A high-pressure gas explosion is a major disaster.

17

Electricity — Friend and Foe

Nearly everyone has experienced at least a slight electric shock at the fingers, which feels at first like a sharp sting, then is followed by temporary numbness until returning blood circulation starts a "pins and needles" sensation. But there was no serious consequence, so the tendency was to laugh it off, maybe with a feeling that you're "immune" to electrical shocks. Such an idea would be a dangerous error. You were lucky in that encounter and should know why the effects were not serious. Never forget that ordinary house current packs a tremendous wallop, capable of killing you or a member of your family.

Electricity is so much a part of our everyday activities, so dependable and easy to use, that we give hardly a thought to the powerful current inside the wires. Much care and strict regulatory controls have been put into the design, construction, and installation of electrical devices, so that even a child without the slightest knowledge about electricity can switch on lights and use certain appliances.

The Shocking Facts. But all too often we come across an item in the local newspaper about a tragic incident in which a person has been electrocuted in his home. Most of the reported fatalities involve men doing construction or maintenance work inside or outside the house, but many victims are women and children. Nearly always, the fatal shock occurred when the victim was using an ungrounded tool or appliance in which, through damage, dampness, or other cause, the casing became charged with live electricity, and the person became the conductor for this current to reach ground.

Here is a brief explanation of what this means: A house electric circuit basically has two wires: one is black, called the lead or "hot" wire, and is connected to the brass terminals on switches and receptacles. The other is the "neutral" or ground wire, with white insulation, and is connected to the light-colored terminal screws of the devices. This neutral wire is grounded directly to the earth, usually through the water pipes, as a protection against lightning and also to avoid the effects of high potential voltage above ground. The polarity of the wires must be maintained throughout the system, that is, all black wires spliced together, and the same with white grounding wires—they are never mixed or crossed.

What Are Short Circuits? Electric shocks sometimes are attributed to a "short circuit." This is not correct, since a short would blow a fuse and disconnect the current. A shock can occur, for example, when the "hot" wire of an appliance or electric tool comes in contact with the metal frame or casing of the appliance; this can happen if the insulation of the cord becomes frayed and the bare wire touches the outside housing, or if the terminal connectors break off, or the inside insulation is damaged. The tool is then said to be "hot," or alive with current. Dampness of the insulators inside the unit also can cause this condition, since water is an electrical conductor. *There is no way you would know about this just by looking*—only an electric discharge meter would show it up!

The human body, since it is so largely composed of water, is an electrical conductor. If you touched an open "hot" wire with one finger, and the ground wire with another finger *of the same hand,* you would receive a slight shock. Current passes along the line of least resistance to the nearest ground, so only the two fingers would be affected. Since the shock was slight, reflex action would pull your hand away, and the only consequence would be momentary local numbness.

But if you were touching the exposed hot wire with one hand, or holding an electric drill in which the casing had become hot, and the other hand grasped a water pipe or sink faucet or heating radiator, the current would surge from one hand to the other, across the chest which is a vital part of the body, with possibly serious effects.

How Tools Become "Hot." That is the fundamental explanation of electrical shocks. But there are other relevant and important details. One is the various ways in which a tool or appliance housing becomes electrified. Another is the various forms of conductive grounds in addition to metal pipes, for example damp concrete or bare earth. And most important to the reader is the means that are available to prevent such tragic accidents by proper grounding of the appliance casings.

A noted television actor and producer, Frank Silvera, was electrocuted while repairing a sink grinder in his home. A Long Island homeowner was killed by electricity in his attic while using a portable circular saw to cut an opening for an exhaust fan. A Clifton, New Jersey, man was killed while using an electric drill to install a downspout bracket on his house wall. A stage and film actor, Donald Clement, was electrocuted while shifting a plugged-in refrigerator in his new apartment. Many children as well as adults have been electrocuted in the bathtub while tuning a radio, or when the radio fell into the water—this latter circumstance differing somewhat from the grounding principle stated above. Many have died using ordinary appliances in the kitchen, bathroom, or backyard. About 1,000 such fatalities occur annually in the United States.

What Are Conductors? There is no effect at all when you touch one exposed wire, even the hot wire, if you are not in contact with either the second wire or a ground conductor. This may be compared with the third rail of an electric railroad or subway line. You often see birds alight unharmed on this third rail. However, any dog or other animal that puts a paw on this third rail while standing on damp earth would be electrocuted. Transit systems use as high as 800 volts in their power feed, but even the 120-volt house current can be lethal.

It would be well to mention here that the low-voltage doorbell and burglar alarm circuits are safe to touch since their transformers reduce the voltage down usually to between 6 and 12 volts. It is interesting also to note that automobile electric systems run on 6 or 12-volt batteries, but touching one of the spark plug wires can give you quite a belt. The reason is that the coil acts opposite to the house transformer, producing a hot spark for the cylinder ignition. When installing low-voltage burglar and fire alarm circuits, the only precaution is to space those wires an adequate distance from the house current circuits, so there is no possibility of a cross current on the low-voltage wires.

Polarized Grounding Receptacles. A big step toward electrical safety was taken when manufacturers of appliances and tools began to equip their products with three-wire cords which include a new type of attachment cap. This has, besides the two regular contact prongs, a round or U-shaped grounding plug. The third wire in the cord is attached inside the tool to the metal housing, and the U-shaped plug that fits into a special type of wall receptacle is separately linked into the grounded house system. Thus a continuous ground would be automatically established when the cord is plugged in, thus retaining continuous polarity. If the tool casing becomes alive with electricity, the current would pass through the ground wire, taking the path of least resistance, and thus protect the holder of the tool.

Another major advance toward electrical safety is in the double-insulated tools with plastic cases which are nonconductive and do not need grounding connections.

Converting the Outlets. All receptacles in the house should be changed to the new polarized grounding type, but priority goes to those in the laundry room, kitchen, basement workshop, garage, all outdoor receptacles on the patio and in the backyard, certainly those connecting any swimming pool equipment and the convenience receptacles used for gardening appliances or tools. The duplex receptacles cost only about 75 cents each; the changeover takes just a few minutes for each receptacle, requiring only about 6 inches of No. 18 wire, stripped at both ends.

After pulling the fuse for that circuit, remove the old receptacle by disconnecting the wires. Attach one end of the new grounding wire securely with a screw to any available part of the metal wall box, and attach the other end of this wire tightly to the green-colored terminal of the new receptacle. Now reconnect the regular current wires, retaining the polarity by connecting the black wire on the brass-color terminal screw, the white insulated wire on the light-colored

Replacing Receptacle

Electric receptacles should be changed from time to time as they lose their spring tension so that contacts are poor and sparking results. Also old types are not adequately grounded. Removing cover plate exposes the two screws that hold the receptacle in the wall box.

After disconnecting the circuit (test it with a lamp or test bulb to be sure current is off), remove the screws.

Draw out the receptacle by straightening the wires to which it is connected. Remove the wires from the terminals and discard the old receptacle.

Modern grounding receptacles have an extra terminal, painted green, for the ground wire at the back. Cut a piece of solid core wire, 14 or 16 gauge, and strip both ends. Wire should be about 6 inches. Crimp a terminal connector on one end so that the wire can be attached easily to a screw inside the metal box.

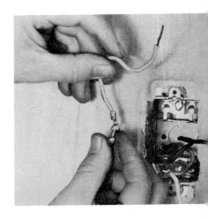

Make sure the ground wire connection is to a screw that goes into the metal box itself so there is good contact. Tighten it firmly. Attach the electric wires to the other terminals of the receptacle checking for polarity; that is, the black wire on the brass screw, white wire on the silver screw. If both wires are black, make the connections and then check them with a test bulb.

The modern grounding plug fitted on most tools and appliances can now be used properly.

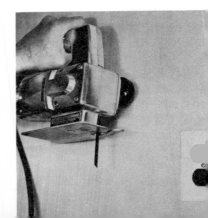

terminal screw. The new receptacle can then be placed into the box and its cover screwed on.

Testing for Continuity. The ground continuity is carried by the metal cable sheathing or conduit, which is attached to the junction box. If the cable is the non-metallic type, and has an uninsulated third wire for grounding, splice and solder this wire to your short grounding wire inside the box, or connect both wire ends together on a single screw inside the box. If the wiring is non-metallic cable and has no third wire, then the grounding wire from the new receptacle must be grounded by attaching to a water pipe or other permanent ground. After each installation, test the grounding with a trouble light consisting of a bulb socket with short wire leads, the ends of which are stripped. After replacing the circuit fuse, insert one probe into one side of the receptacle, and touch the other to the grounding contact. If the bulb does not light, try the first probe in the other side of the outlet. If neither side lights the bulb (but the bulb lights up when the stripped test wires are placed into both current sides) then the grounding connection is faulty and should be checked through until contact is made.

Grounded Extension Cords. Putting an ordinary extension cord in line with your grounded tools cancels out the grounding protection. Instead, make up several three-conductor extension cords, of the different lengths that you generally use, so that you will have them on hand when needed. The special fittings that are required for these cords are available at most electrical supply houses, or from Sears Roebuck (Catalogue No. 34 P.5927 and 5932, $1.62). Ready-made

Grounding is essential for an outdoor receptacle. Use of the adapter shown here is makeshift and not fully reliable. If used, pigtail ground lead, normally green, must be firmly attached to the cover screw.

cords in 25, 50, and 100-foot lengths and in 12, 14, and 16 wire gauges, also are available from Sears.

Variables in Conductors. The severity of an electric shock depends on the line voltage, the course of current travel through the body, and the physical condition of the person. The condition of the skin also offers varying resistance to passage of the current—a wet, greasy, or dirty hand conducts current more readily, while a thick and dry skin offers greater resistance. The latter condition accounts for the fact that manual workers are less subject to shocks than others, since calloused hands offer some insulation.

Any electric tool used outdoors, in the garage, or any place where moisture is present, spells DANGER in capital letters, unless the tool is properly grounded, or has double insulation. This is proven by the frequency of news items reporting serious consequences in such circumstances. One such fatal incident involved a homeowner who was connecting a sump pump to drain his flooded basement after a rainstorm. Another tragic accident occurred when a youth, after washing down and waxing the family car in the driveway, picked up a drill which was fitted with a lambswool disk to buff the wax. The hose spray possibly had dampened the insulation inside the drill, or the boy's feet and shoes had become soaked when sloshing through the puddles left by the garden hose. Similar fatal consequences followed use of ungrounded tools on wet grass, on damp earth, even on garage concrete soaked with grease and water. All that was necessary to avoid these tragedies was to use the grounding protection now provided with the better quality tools or equipment.

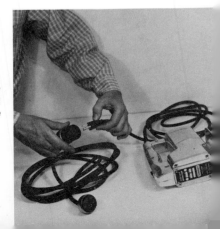

When using an extension cord, especially outside or near a possible shock hazard like a water or heater pipe, use a 3-wire grounding cord.

Double-Insulated Tools. Because many homes still do not have the safe-grounded electric receptacles, and perhaps in recognition that many homeowners are neglectful in following up on grounded connections and extension cords, manufacturers now produce tools with double-insulated lead wire connections and motors and plastic housings. These provide the necessary extra measure of safety. Even so, it is a good idea to become fully aware of the grounding principle and follow through, particularly with equipment used outdoors or where there may be an inadvertent ground contact.

The Kitchen Sink. Here lurk many electrical hazards because of the ever-present water faucets. All appliances, like toasters, blenders, mixers, and broilers, should be kept a sufficient distance from the sink so that there's no possibility of touching both the appliance and a faucet handle at the same time. Only those appliances specifically stated to be immersible should be immersed in water. An appliance must always be disconnected before cleaning, and never allowed to become damp. If an electric shaver drops into a filled sink bowl, *DON'T REACH FOR IT* until you've unplugged the cord. Then, even though the shaver may have a plastic housing, do not use it until you're sure that its electrical integrity remains intact and the motor has been thoroughly dried. Hair dryers and electric heaters are additional sources of serious accidents in the bathroom, where the user may be standing barefoot on a damp floor, or attempt to handle the appliance with wet, soapy hands at the sink.

Some people think that the electric current is disconnected when the appliance switch is "off." Not so. As in the case of a radio, the current is right there even if the switch is off, and a shock can result if the radio becomes wet or falls into the water. The emphasis is made again: always pull the plug.

Electricity at the Swimming Pool. Haphazard wiring and improper equipment constitute a serious threat, particularly at surface pools which have the filter system located directly at poolside. Wiring often consists of ordinary electric cord laid along the ground, where the insulation is likely to become frayed and soaked. The area around the pool is always wet and bathers likely to be barefoot, thus any contact with a damaged electric cord can result in a serious shock. Filter cords should be of the 3-wire, grounded heavy duty, plastic insulated type, specifically rated for use in damp locations. The Sears No. 34 P. 5817 cord, 14 gauge, is suggested for this installation. The receptacle should be located by cable wiring as closely as possible to the pool, keeping the connecting cord short. If possible, string the cord overhead to the filter motor, otherwise arrange a wood-encased channel for protecting it. If there is any doubt about the grounding connection to the house water system, drive a copper-coated 8-foot-long rod all the way into the ground, and attach a grounding wire with an approved clamp. It is not good practice to suspend lights above the pool, since winds or other cause may snap the wire or drop one of the lamps into the water. Instead, mount flood lights securely to tree branches, poles, or nearby parts of the house. In sub-surface pools, underwater lights must be specifically designed and rated for that purpose, properly installed and as with all electrical equipment, bear the U.L. rating seal.

LIGHTNING

Among nature's theatrics, perhaps the most awesome and mysterious is a lightning bolt flashing out of the sky, with the concentrated light of a million electric bulbs. A lightning stroke lasts only about a thousandth of a second, but that moment is sufficient to strike terror in the minds of many. Throughout man's history, even until not many years ago and despite the growth of scientific knowledge regarding its cause and source, numerous cults regarded lightning as the weapon of wrathful gods. Human victims of lightning were considered punished for angering the heavens.

Lightning kills 500 persons annually in the United States. Of these, 400 are men, 100 women and girls – obviously because more males work or play out in the open. Many of the victims are struck while engaged in some sport activities – swimming, fishing in small boats, or golfing. Frequent reports are seen in the press of picnickers, caught in a clearing when a storm started, struck after making a dash for shelter under

the nearest tree. Some of the deaths were of workers on various construction projects in the open. Rarely, if at all, do fatalities from lightning occur in homes or buildings.

The safest place to be when lightning strikes is inside a building with a steel frame, such as a typical office building. The Empire State Building is struck many times each year, because of its height, without damage or injury to any person. Homes in the city also are quite safe since they are usually surrounded by taller buildings or structures like transmission line pylons, or bridge towers. Homes also have their own built-in protection because of their extensive metal installations, such as the plumbing system, which are grounded to the central water supply lines. Country homes of frame construction are more vulnerable, but can be protected with lightning rods. This is a metal rod starting at the topmost part of the house, usually just above the chimney, and connected to a copper-coated pipe driven at least eight feet into moist ground. The connections between the rod and ground electrode are required by the Electrical Code to be run in as straight a line as is practicable. A house with standard electric system, which itself is always adequately grounded, would not need a separate lightning rod.

These are the facts of our present times, but until the middle 18th Century, lightning was a subject of considerable concern and with good reason—aside from the fears of this natural phenomenon. Lightning plagued mankind, caused many fires of homes and barns, and many deaths since people were not schooled in correct preventive measures. Superstition, in fact, led many into actions that were disastrous.

Went Wrong Way. It is recorded that the peasants in some parts of France were convinced that bell ringing would ward off lightning, and rushed at the start of a storm into the tall bell towers, which were precisely the most likely spots for lightning strikes, since the towers were the tallest structures and were entirely of stone, without any conductive metal. The French Parliament eventually banned the practice, but it continued for a long time nonetheless. Similarly, the custom of storing munitions in vaults beneath churches resulted in thousands of deaths when the towers were struck by lightning and the gunpowder detonated.

Lightning also started fires on many ships at sea, so that many a warship was blown up when a lightning bolt set off the ammunition. That was before the days of steel vessels; now such events could not occur with metal-hulled ships. More than 10,000 forest fires are started each year by lightning in the United States—present day fire-fighting facilities, including the use of planes, limit the damage. Even so, millions of acres of woodland are destroyed annually.

As everyone knows, the big change was brought about when Benjamin Franklin demonstrated with his kite that lightning was the same electricity as that produced by the hand-cranked magnetoes of that time. His experiments resulted in the development of the lightning rod, of which the first one was installed by Franklin in 1752. The description of this device in Poor Richard's Almanac resulted in universal application, and soon the lightning rod was in almost all homes.

Outdoor Television Antennas. These are usually located so that they extend above the level of the roof, sometimes above chimney height, and thus would be the focal point attracting lightning flashes, or static charges from thunder clouds. The Electrical Code requires that antenna lead-in connections include a lightning arrester. This is a small ceramic block, usually attached to the exterior house wall with a ground lead from the arrester attached to a cold water pipe, or to a copper or galvanized ground rod driven into the earth. Another method of grounding that is used occasionally is to attach a wire from the antenna directly to a vent pipe; this method does not provide maximum safety since it is generally unwise to run static-discharge conductors directly through the house before they are grounded.

What happens when lightning strikes an automobile? The lightning bounces through the steel frame of the car until it is used up, since the car is on tires and not grounded. The passengers will be unharmed if they remain inside, protected by the metal shell in which they are enclosed. There is an unanswered question about what would happen should a person step from the car at the precise instant of the lightning bolt. There is no known instance to verify any conclusion about this—with all the millions of automobiles always on the road in the worst of storms no reports of injury to occupants from such situations have been recorded, or verified. Airplanes, of course are not grounded; their metal envelope absorbs and dissipates the electrical energy of the bolt. But this is not true of lighter-than-air dirigibles, and a French hydrogen-filled airship was destroyed when hit by lightning.

18

Safe at Home

The home is rightfully expected to be a safe haven for all the family. The requirements to achieve this are sound building design, adequate maintenance, elimination of possible hazards, safe products, and adherence to sensible accident prevention rules.

Wide breaches in this protective shield are evident in the regularly released government statistics: more than half of all accidents that cause some degree of disablement occur in or about the home. While the 27,000 fatal home accidents in 1969 were substantially fewer than the previous year, there were increased deaths from fires, poisoning, and misuse of firearms.

A wise move for every homeowner is to make periodic hunts for potential hazards. The check list at the end of this chapter may prove very helpful.

Ceilings Falling. Homeowners sometimes are unaware of the personal hazards that often accompany structural defects or physical deterioration of the home. A typical example is the collapse of a ceiling. Plaster crumbles when wet and can lose its bond to the supporting surface. Water from a leaking roof, an overflowing bathtub or stall shower pan on an upper floor, or rain blown in around an uncaulked window frame, seeps down and soaks the ceiling underneath. The plaster becomes detached from its lath and crashes down suddenly in huge chunks. Since plaster is extremely heavy, such a ceiling collapse represents a distinct physical danger in addition to the considerable repair bill. Usually, plaster that has become damp shows a tell-tale stain; do not paint over this stain until the ceiling area has been tested to make sure that it is firmly bonded, and the source of the moisture located and corrected.

Falls. Approximately one third of home injuries result from falls; these may be falls from ladders, a roof, tree limb, an unsteady step stool, or on stairs that are poorly lighted or have frayed carpeting or tread covers. Many serious falls result from tripping over an encumbrance, or some defect in the floor surface such as a cracked threshold, loose brick in outside steps, or uneven sidewalk flagging.

Falls may result from a structural defect, as was emphasized by one tragic incident in which 25 persons were injured, several of them critically, by the collapse of a 15-foot-high wood patio deck during an outdoor cookout. The party was attended by 40 persons who crowded the deck, which was built onto a new home, level with the dining room. The raised patio, which obviously had been constructed without adequate supporting piers and with deck boards of insufficient thickness, gave way under the exceptional weight and sent the party guests tumbling in the equivalent of a two-story fall. Many of the victims were also burned when kerosene used in torches for outdoor lighting spilled and started a flash fire in the wreckage.

All outdoor wood floors are subject to rapid deterioration unless properly maintained. Painting each spring and autumn with deck enamel may be necessary in certain localities to preserve the wood. The underside of the floor planks should be coated with a good deck varnish at least annually, and examined regularly for signs of wood rot or other deterioration. Heaving of the footings during exceptional frosts may throw supporting piers or columns out of plumb. These should be repaired or rebuilt promptly, and any cracked or rotted floor planks replaced.

Decks, steps and walks are safe when the base is solid enough to provide firm footing, the surface without any snagging irregularities, smooth but not slippery. Examine stair railings to be sure they are securely anchored, reinforce when necessary, and paint metal railings at regular intervals to prevent destructive rusting.

Look Out Below! A reverse kind of fall can be just as painful. That is when heavy objects come crashing down. A hammer or other tool left on top of a stepladder can produce serious injury when the ladder is tipped. Another all-too-frequent event of similar nature is careless placement of heavy objects on the edge of high shelves, out of sight but easily dislodged by anyone reaching into a closet.

Overloaded shelves, particularly the type supported on wall hung standards, can collapse with the massive accumulation of record player, audio system, and record albums — these shelves are frequently seen sagging at the center from the excessive weight and can come crashing down at any time.

Much of this trouble derives from the current preference for suspended wall shelving which has a more attractive "open look" than the old-style solid bookshelves that are supported with uprights resting on the floor. The wall standards used for suspended shelving should be attached securely with screws of adequate size and should be spaced no farther apart than 30 inches for uniformly distributed loads such as books. When the shelves are to carry heavy audio equipment and collections of record albums, space the standards 24 inches or less apart. Width of the shelf stock is a factor, too; shelves more than 10 inches wide require strong support members.

See-Through Patio Doors. A comparatively new hazard has developed with the growing popularity of sliding glass patio doors. The large glass panels are not always clearly visible, particularly to persons with impaired vision, and numerous accidents have occurred. Some of the glass is so thin and poorly framed that it shatters when anyone merely walks into it. Children have been severely injured by the broken glass when they tumbled against the doors while playing. In one serious incident, a noted singer slipped on the wet tiles by the swimming pool of her Bel-Air home and fell through the glass, suffering extensive cuts that required several hours of surgery.

Recommended safety measures include: 1. Glazing the doors with heavier glass, preferably ¼-inch tempered safety plate, or ½-inch insulating glass. Replacement of the glass with sheets of clear plastic like G.E. Lexan is a possible solution which has not yet been fully developed. 2. Placing decorative decals or tapes at eye level so that the glass is easily spotted, and 3. Seeing that all floor surfaces in the door area are slipproof.

Handling Power Mowers. Millions of power mowers are purchased each year by eager homeowners who see a release from the wearying chore of pushing a hand mower around. The gasoline or electric-powered mowers are indeed a boon — just guide it around and it does all the work itself. But the mower doesn't stop only at cutting grass, but will take a finger or toe along if it gets into its clutches. It has been estimated that power mowers cause over 100,000 injuries annually. So while the mower now does the work, you must turn your attention to controlling it to prevent accidents.

The most common injuries occur as a result of trying to clear the discharge part while the motor is running. Since the blade turns so fast that it is invisible, there's far too much chance of your fingers getting caught and sliced. Instead, turn off the engine each time before clearing the chute. When mowing on a slope, or wet grass, where a foot may slip under the safety guard and into the blade, be extra cautious. Always wear suitable shoes — golf shoes are excellent for this.

Using a power mower on a steep slope presents a problem. With a push-type mower, the cutting swath should be lateral or side to side, while in the case of a riding mower, the cut

Thorough operating knowledge of your power mower is the first step on the road to a happy and safe mowing season. The Outdoor Power Equipment Institute urges all power mower owners to read the operator's manual and safety rules included with every unit. For the best and safest mowing results, manuals should be carefully reviewed each year.

is made vertically, or up and down. Power mowers should always be pushed, never pulled.

The same cautious approach is necessary also when using snow blowers, garden tractors, tillers and all other power equipment around the home.

Here are the 10 safety rules recommended by the Outdoor Power Equipment Institute:

1. Study the manufacturer's manual thoughtfully and do not operate the equipment until you are familiar with the correct procedures. In the final analysis, safety results from the operator's knowledge of the equipment and its application.

2. Rake the lawn occasionally to remove small stones, twigs, and other foreign objects that can be tossed by the blade or clog the ejection ports.

3. Clear children and pets from the area. Be watchful and stop the machine if any return.

4. Keep hands and feet clear of the discharge port and blade.

5. Wear proper clothing, not shorts or sandals, or go barefoot.

307

6. Stop the engine and disconnect the sparkplug before making any repairs or adjustments. Don't refuel a hot or running engine.
7. Never allow children to operate a power mower, or let adults run the mower without adequate instructions.
8. Maintain the mower in top condition.
9. Don't carry riders while using a riding mower.
10. Always look behind when backing a riding mower.

Pool Casualties. In a period of just 20 years, the number of permanent in-ground swimming pools in private homes zoomed from only 2,500 to more than half a million by 1966, and an estimated 750,000 by 1971. This amazing growth was outpaced only by the placement of several million portable on-surface pools, plus the usual complement of wading pools for small children. Unfortunately, the mushrooming growth in the number of pools—and the number of persons for whom the use of pools for swimming has become available—has exposed the younger population to the great menace of accidental drowning. The number of drowning fatalities, surprisingly, is not accurately known; only in recent years has the Federal government initiated procedures to assemble this data from the only available source, by scanning local newspapers from all over the country. The first figure obtained was for 1965, showing 484 pool fatalities, of which children under age 5 accounted for almost half (43 per cent). The obvious inference is that the drownings resulted from falling or slipping into the water. Studies showed that the most frequent contributing cause of drowning appeared to be the lack or inadequacy of adult supervision. In drownings of older children and adults, trespass (unauthorized use of the pool without supervision) was believed the greatest contributing factor.

These statistics and the study conclusions are cited to reinforce the recommendations by various authoritative sources:
1. Adequate enclosure of the pool area, with self-latching gates and fences built both close enough to the ground so that infants could not crawl beneath, and high enough (at least 6 feet) so youngsters would have difficulty scaling the top. A particular hazard was found to exist where the home

serves as a side of the enclosure and has sliding doors that give direct access to the area.

2. Wading pools should be emptied when not in use, or fully fenced.

3. Competent adult supervision at all times that the pool is in use. Pool owners should seek instruction, usually available from Red Cross or YMCA units, on safety practices and the essential skills for safe pool operation.

4. All electrical installations, lighting, and accessories should conform to all Code requirements, properly grounded, frequently inspected and with emergency night illumination available. See Chapter 17.

5. Safe conduct practices in and near the pool established for all users. Purity of the water and sanitation principles maintained. Rescue devices (a long pole, float, buoys and floating line) always accessible.

High Pressure Gas. Natural gas is piped into city homes under high pressure, then a reducing valve lowers the pressure before the gas enters the meter. The installation must conform to strict utility regulations, is inspected periodically, and should be completely trouble-free if protected from damage. Occurrences of gas escape from the high pressure source are rare, but when anything does happen with the high pressure line, the effects can be disastrous. A series of such events, however, has aroused fears of further gas explosions, and pointed up the public's lack of knowledge regarding this hazard.

If the meter is inside the house, the gas installation includes a shutoff valve at the pressure reducer. There is another shutoff valve in the street in front of the house that can be reached by utility crews in an emergency. The homeowner should learn the location of the gas service pipe inside the house — usually in the basement — and study the situation to see whether there is any possibility of inadvertent damage to that pipe. If the installation is close to the bottom of a staircase, decide whether anything carried down the stairs such as a piece of furniture or a ladder might bump into the pipe with sufficient force to break off the valve cap. In one historic incident an upright piano being carried down the basement stairs broke loose and slid down, smashing into the gas main and causing a devastating explosion. If the valve is in a vulnerable position, be sure to build an adequate protective shield around it. Additionally, always maintain a healthy regard for the gas installations, following these suggestions:

1. Locate the street valve cover, keep it all times swept clean and clear of obstructions (such as plantings). In winter, keep the valve cover always clear of snow and ice.
2. Learn how to turn the gas valve at the house meter, but avoid unnecessary tampering with this valve or other gas apparatus.
3. Have gas appliances installed with the proper pipes and connecting hoses — never use substitute or makeshift materials. Any repairs or maintenance to be done by specialists experienced in servicing gas appliances.
4. Keep pilot lights correctly adjusted, clean and lighted.
5. At any indication of a gas leak (other than from a pilot light) have all occupants leave the house and notify the gas company emergency service.

Electric tools were never designed to be carried like this, by their cords. Such treatment can rip the cord ends off their terminal connections — then when you plug in the cord, there is live juice running through unattached wires.

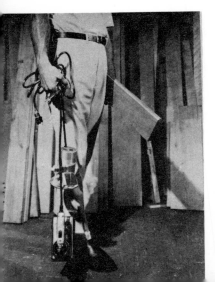

First principle when working with any power tool is to get the hands out of the way. The work should be held securely in a vise, or other safe means. The saw blade, also, should be of proper size for the job at hand.

Cautions on Power Tools. It is reasonable to regard as part of the house the tools required to maintain it in condition. The modern home has grown so complex, has so many details to keep in working order, that the task cannot be done amply with simple hand tools. The array of power tools available today is indeed essential, performing tasks that otherwise would require an extensive range of skills and considerable time. There are, in addition to the circular saw and electric drill, which may be considered the basic tools, such useful instruments as routers, saber saws, hedge clippers, lamination trimmers, planers, mowers, lawn tractors, chain saws for cutting trees and logs, and dozens of others that seem to complement the growing list of kitchen electrical accessories. Unlike the well-safeguarded kitchen appliances, however, electric tools can be extremely hazardous when used carelessly, or by a person unfamiliar with the technique involved.

Bench saws and portable circular saws generally earn careful respect if only because of the shrill sound emitted by these saws. But even an experienced woodworker must exercise extreme care every moment when using these tools — clothing must be of correct type to prevent any entanglement with the moving blades; the fence precisely aligned so the wood stock will not bind; the blade always good and sharp to obviate excessive—and dangerous—application of force to the work; the stock held at the right position; work fed through with a push stick where required; protective guards in place.

These injunctions are not stated to discourage use by anyone of the saws or other power tools. On the contrary, the view is emphasized here that almost everyone can and should be able to operate power tools. There are over 100 million licensed drivers of automobiles in this country—requiring a far greater skill than needed to operate any power tool safely.

What is essential, however, is that the tool user obtain some skill or training beforehand regarding each tool, that the proper tool be selected for the job, that it be maintained in safe condition, and handled with adequate care.

Whatever electrical tool is used, make certain that it is electrically grounded, including use of three-wire extension cords with continuous polarity, or that it is double insulated with all-plastic housings. (A complete statement on causes of electric shocks, and the methods for grounding appliances, is given in Chapter 7.)

Tool manufacturers certainly want to sell their products, but they are definitely concerned also that the tools be used safely and effectively. Many manufacturers are members of the Power Tool Institute which conducts continued consumer education programs, and the instruction booklets supplied with most power tools contain valuable information regarding use of the tool which should be studied carefully.

Storage of Chemicals. Many homeowners lack adequate knowledge about safe handling and storage of household chemicals, particularly paints and their solvents, cleaning compounds other than detergents, and insecticides.

Some paints are highly flammable, even explosive. Others are not. Latex and masonry paints, for which water is the "solvent," are non-flammable and generally non-toxic, may be safely stored in sealed containers in the home.

A neat well-organized workshop is a safe workshop.

All oil paints, varnishes, stains and lacquers are flammable, as they contain mineral spirits and linseed oil, or petroleum distillates, alcohol, lacquer thinners, and other highly volatile thinners. These paints must be kept in tightly sealed metal containers. Only a small quantity should be stored in the home, always in a ventilated cupboard. The various solvents, including benzene, varnoline, alcohol, lacquer thinner, and contact cement thinner are even more volatile, and must be carefully stored. Keep supplies on hand to a minimum at any time, always in metal containers with tightly sealed caps. Avoid glass containers, as accidental breakage would flood an area with a large amount of the fluid that could cause an uncontrollable flash fire.

19

Nature's Perils to
Your Home

Neither rampaging forest fires, nor raging floods, nor devastating earthquakes, not even breath-balking smog, has restrained the growth of California—our fastest-growing state. The harrowing experiences in the 1970-71 sequence of natural disasters in Southern California conclusively demonstrated that even such a one-two-three punch could not shake the attachment of these residents to their homes.

First they fought off the biggest brush and timber fires in the state's history, when the enormous blaze that swept in from the Mohave Desert at speeds approaching 100 miles an hour destroyed or burned more than 1,000 homes, forced the evacuation of 50,000 residents from small communities in the area, and scorched hundreds of square miles of forest lands.

Soon afterward came the floods, their torrents overflowing river banks, gouging out the land and undermining foundations of homes, strewing mud and rocks broadside, washing out culverts and heaving up stretches of roads.

Allowing just long enough for matters to settle down a bit,

Raging winds of Kitchen Creek brush fire in California carried burning embers which ignited roofs and destroyed many homes at a distance from the fire itself.

the ordeal of earthquakes began early the next year, collapsing even newly constructed buildings of reinforced concrete, blocking the freeways, and threatening to burst a huge dam at the head of a densely populated valley.

"Something like this makes you realize that you can't buck nature. But now that we've sweated it out, we feel that the worst is over, and there's no reason to leave." That explanatory remark by one family sums up quite succinctly the prevalent attitude as residents worked to restore their homes and resumed their everyday pursuits. Few families pulled up stakes, perhaps because they felt there was no place else to go, perhaps because they felt rooted with their homes and were determined to remain, come what may.

But will they act on the dire warnings of scientists and prepare defenses against the predicted greater onslaughts of nature's forces? Or will they persist in some fatalistic notion that nature's upheavals are unpredictable, so it's best "not to worry" but instead trust in faith?

Justification can be cited for both views, of course. The hard scientific facts, however, cannot be simply brushed away. A noted physicist predicts that a major quake, equal to or greater than those at San Francisco in 1906 and north of Los Angeles in 1857, will occur before the end of the century, most probably originating between San Francisco and Los Angeles, and that the massive quake could wipe out both cities causing a million fatalities if it takes place during working hours, unless adequate precautions are taken. "Where there have been earthquakes in the past, there will almost certainly be earthquakes again," is a scientific axiom.

The terrible brush fires are inevitable in Southern California, where three conditions exist: Vast areas of explosively flammable chaparral plants: the hot dry gusts of the Santa Ana "devil winds" that fan and spread the flames, and insufficiency of water to curb the wildfire.

And there will again be the torrents which periodically tumble down the mountainsides, flooding the drainage basins, washing out culverts, converging into the canyons, smashing entire colonies of homes.

Earthquakes are not endemic to California alone . . . they can occur again in Boston or Charleston, S.C. which had suffered quakes before, or even possibly in New York or Houston. Scientists cannot predict when, where, or with what force a quake will strike. But there is special reason for concern about the California coast because of evidence that strain has been accumulating along parts of the San Andreas Fault that runs from Northern California to Mexico. One authority, Dr. Jerry P. Eaton, head of the National Center for Earthquake Research at Menlo Park, Cal., believes that the 1970 California earthquake may have transferred strain from a small fault to the main San Andreas Fault, bringing nearer the day of the great quake.

Various scientific specialists are convinced that California is unprepared for this eventuality. Study reports by two Federal agencies point out that thousands of buildings are "ripe for collapse" and that even modern buildings constructed to supposedly earthquake-resistant designs were vulnerable, as shown by the fact that practically every reinforced concrete column and some exterior beams cracked during the 1970 quake. The Presidential Task Force on Earthquake Hazard Reduction declared its concern regarding old and vulnerable buildings, particularly the multi-storied buildings erected several decades ago. It pinpointed as the greatest threat the existence of unreinforced brick parapets, stating that more persons would be killed by the falling masonry than would perish in collapsing buildings. Adding to the hazards are the water storage dams built above populated valleys, the increasing number of high-rise buildings, and the area's dependence on freeways which could become completely blocked by the fall of any bridges.

Few specific proposals for quake protection have been offered by the scientific authorities. Their contributions are important if they arouse the area to the need for action, though there still may be doubts as to the proper forms that the action must take. An alerted community will find a way to its salvation. Perhaps in the coming years, science will discover how to foretell accurately the time and intensity of a coming shock. And also, perhaps, the best defense that can be made is to develop a sure and quick means of evacuating a threatened population to wait out the shock.

FOREST FIRES

Homes in deeply forested areas are vulnerable, and every possible effort should be made to reduce the fire hazard. Chief protective measures are landscape clearing and brush cutting to eliminate trees or bushes close to buildings, selection of plants that do not burn readily, asbestos house siding, tile roofs, and other fire resistant building materials, and adequate water supply with sprinklers.

The Agricultural Extension Service of the University of California recommends wide spacing of flammable native shrubs, and their close pruning to remove old quick-burning growth, as important aids in minimizing fire hazard. California law requires complete brush clearance within a mini-

In forested areas, landscaping must take account of forest fire danger. Trees and brush around the property should be cleared back. Grass or other ground cover is important. Plantings on the property should be widely separated. A fireproof roof is a sound investment in such an area.

mum of 30 feet around all structures, while a 70-foot rule applies in areas of extra hazard, where vegetation must be kept at less than 18 inches in height. Chaparral brush should be removed and replaced by low ground cover. Unwatered chaparral is regarded as especially dangerous when it is growing densely in a canyon below a house in which the location creates a chimney draft situation if a fire should occur.

Additional measures are planting of grass lawns or ground covers around all structures, and most important, keeping road areas cleared so that retreat will not be cut off, if it becomes necessary.

WINDSTORM

The Dread Twister. Roaring like a hundred thunders and with its black, funnel-shaped cloud whirling at a speed up to 500 miles an hour, a tornado usually destroys everything in its path, lifting entire buildings from their foundations and smashing them into rubble, tearing huge trees right out of the ground, sweeping bridges off their piers, tossing freight trains and locomotives around like toy boxes.

Tornadoes occur usually during late Spring and Summer,

There she blows: Twister's funnel can appear suddenly, without warning, accompanied by shrill noise like that of plane jets. Black funnel-shaped cloud spins violently around a center core, often at speeds up to 500 miles an hour.

most often in the central and southern regions of the United States, but they can appear almost anywhere and at any time of the year. Kansas and Iowa have experienced many tornadoes, but these storms have hit every state in the union (except Hawaii, which copes with typhoons). Major cities as well as the open countryside experience these ravaging storms, but the twisters are rarely seen over mountains or very dry regions. In a single day in April 1965, a series of tornadoes over Midwestern states destroyed nearly 8,000 homes.

Simply stated, a tornado is a tightly compressed mass of air spinning at fantastic speed around a low pressure core. The turbulence begins high in the clouds and extends downward toward the warmer air below, dragging along bits of cloud into the whirling mass. The surging cool air also condenses moisture as it whirls and forms into the funnel shape, producing the characteristic dark appearance. The tornado may be quite small in cross-section — anything from a few yards to about a quarter of a mile — and may travel some 10 to 20 miles in distance from its point of formation. The storm usually is followed by a heavy rainfall.

No Early Warning. Twisters form so quickly and unexpectedly, sometimes developing in what moments before had been a clear sky, that no reliable predictions of their appearance can be made. The warning is by sight or ear— the black funnel shape can be seen from afar on open fields, leaving just enough time to get into a shelter, since the tornado travels rather slowly at 50 miles an hour, or the thunderous roar can attract sufficient attention (though the many sounds of the modern industrial and technological age may confuse the listener).

Tornado Shelters. Families in tornado territory have learned the value of having a conveniently accessible storm cellar that can be reached in the few minutes available after a twister is sighted. Suitable for a shelter is an underground place, like the old-time cold-storage cellars, but preferably one that is not directly under the house unless there is an exceptionally solid roof to protect the cellar from falling debris if the house is demolished by the storm. Solid-looking buildings, even stone-constructed smoke houses, are not necessarily safe. Huge 12-story buildings have been lifted right off their foundations by tornadoes.

Lacking a suitable cellar, it is not very difficult to dig an

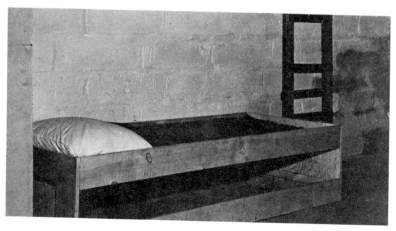

Cyclone shelter can be constructed in the basement of a well-built home. Accommodations can be crude but may as well be comfortable.

underground shelter on the property, but not too near the
house or possible falling objects like a huge tree, and on high
enough ground to assure against flooding the interior during a
downpour. The room need not be at all fancy, but it should
have walls lined with mortared stone, concrete blocks or similar masonry, and an adequately strong roof supported with
steel beams or timbers to guard against collapse.

The doors should open inward so they would not be blocked
in the event debris were piled against the opening, and some
secondary means of ventilation provided. While the storm can
be expected to pass over in short order, there is some possibility of being trapped in the shelter for a day or more by a heavy

Cyclone shelter in basement should have means of exit other than
through house. Here covering over exit steps is supported with
heavy timber. Doors opening inward help avoid entrapment.

downpour. The shelter should be stocked with emergency equipment and supplies — canned food that does not have to be cooked, bottled water, lanterns, first aid supplies, and some sanitary facilities. A battery-operated radio also is an essential item so you can keep informed of local conditions.

Beating the Storm. Lacking a nearby shelter, what can you do if a tornado is heard or sighted in the distance, coming at the speed of 50 m.p.h.? There won't be enough time to reach a shelter, and you can't run for it. You might be able to outpace the twister in an auto if there's enough advance time, and a good clear road to keep going. But if it looks like you can't make it, don't stay in the car — that would be the most dangerous place to be in. Your chances are better if you leave the car and race as far away from it as possible, then sprawl flat in a ditch.

On a field, if you've got to try ducking a twister, your best chance is to run at right angles to the northeasterly path of the storm, that is, either southeast or southwest. Get into any available ditch or depression in the ground, the lower down the better.

For those without provision of a shelter, or caught in a storm away from their cellar, there's no choice but to sit it out in a basement, huddled into the southeast corner to withstand the storm's first blow and remaining close to a wall to avoid any debris that may crash down from above.

Home Protection. If your home is some distance from the town center so that it would take too long to reach the community storm shelter, or there's a possibility that the roads may become impassable during a heavy storm, standing preparations can be made for greater safety during any storm. An emergency food supply should be kept on hand (in a separate place so you know it's always there) as food shipments may be cut off for long periods. Large plastic water bottles, tightly sealed, can be kept for long periods, supplemented with extra supplies drawn immediately at the warning of a major storm. Other preparations include removing tree limbs that could fall on the house if broken off, and installing local electric wires underground.

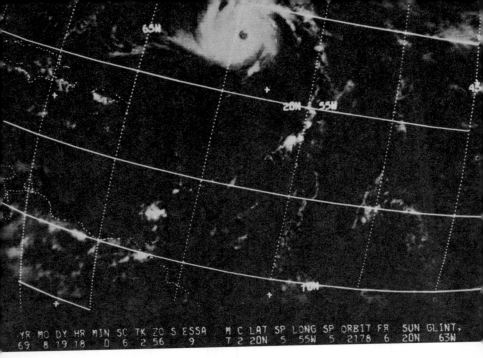

YR MO DY HR MIN SC TK ZO S ESSA   M C LAT SP LONG SP ORBIT FR SUN GLINT,
69 8 19 18 0 6 2 56   9   7 2 20N 5 55W 5 2178 6   20N   63W

The "eye" of the hurricane, measured on photograph taken by NOAA's storm data information service, using B-57 planes for on-the-spot observations.

The tendency when a storm breaks is to close up the house tightly, but this only increases the danger. A window or ventilators should be left open to equalize the pressure and thus reduce the suction action of a low pressure storm.

The Big Blow. Early hurricane warnings made possible by weatherplane patrols and constant observation by satellite have enabled communities to survive what at one time would have meant sure disaster. Some hurricanes are indeed terrifying in their impact and destructiveness; with 24-hour advance notice, it's at least possible to "batten down the hatches." The winds that may reach a force of 100 miles an hour do heavy damage, but the storm fatalities more often are caused by the accompanying heavy rains and subsequent flooding than by the high winds. Communities in hurricane-threatened areas, have the responsibility to provide for their residents' safety in a disaster situation. Committees can be set up consisting of personnel with specialized or technical skills to coordinate emergency plans for police, fire, communications, medical and hospital services, public shelters, rescue missions, water supply, sanitary facilities, food supply, and others.

One or more suitable buildings of stout construction can be selected for emergency shelters, and stocked with the necessary equipment, food, water and medical supplies. An auxiliary power plant is essential, as is a system of warning sirens or other rapid means of notification to residents.

Recommended actions that may be taken to lessen the hazards of severe storms include the banning of overhead power lines, to prevent the swirling of live wires downed by the storm, and elimination of billboards that are extremely susceptible to destruction by storms. The wood planks and metal panels of the signs are blown helter skelter across the landscape like guillotine blades. Huge business signs atop buildings can come toppling down, adding to the havoc in the streets below. One place not to be in during a low pressure storm is near large store windows which more than likely will be blown out, the glass shards spraying everyone in the area.

The gulf coast of Florida once was tagged with the name Hurricane Alley because a series of the worst hurricanes in the United States occurred there in the late 1920s and early 1930s, roaring up the Atlantic coast as far as New England before being spent. However, such storms may originate at any tropical area.

In areas possibly subject to hurricane or gale winds, homes should not be built so close to the shore that storm-lashed waves can reach them, and they should be especially well anchored to their foundations. Masonry homes of low profile have a better chance of escaping damage.

HOME HAZARDS CHECKLIST

Stairs: Treads securely wedged; no loose or torn carpeting; handrails solidly secured; steps unencumbered and well-lighted, with handy light switches at head and foot of stairs.
Snow Slide: From roof. Install slide retarders.
Loose Ceiling: Look for stains.

Water Temperature Control in Shower: Check also mixing valve at water heater.

Stored Paints and Chemicals. Also eliminate oil rags.

Low Overhead Pipes: Paint light color for visibility; build barrier if necessary.

Toys and Other Floor Encumbrances: Always cleared away where they may cause falls.

Retaining Walls: Watch that weep holes are open to prevent washout and pressure buildup at back; repair immediately at sign of tipping or sagging wall.

Low Casement Windows: When open, they cause head injuries; plant bushes or other barriers at those locations.

Window Air Conditioners: They must be well supported on the outside wall to prevent dropping out.

Outside Wiring: Underground of approved type; no exposed wires on ground surface, none in wet areas near pool, all exterior lights with approved switches, preferably remote control.

Garage: Never run car with door closed, or use gasoline to wash parts in a closed garage.

Ladders: Always on wall racks, preferably chained; keep ladders under cover, check condition before using — rungs tight, side rails solid, no sign of split; place ladder so base is a distance from the wall one-fourth that of the length, e.g., a 5-foot base for a 20-foot ladder.

Fire and Electricity Hazards: See the respective chapters on those subjects; but, to repeat, never touch electric appliances (like a radio or shaver) while in the bathtub.

Power Tools: Prevent unauthorized or unpermitted use by locking portable power tools in cabinet; stationary power tools should have key-controlled switches, kept locked when not in use.

Heating Plant: Never use steam radiator without the vent valve; when replacing valve, be sure that threads have matched and that the valve is turned on securely; steam boilers and hot water heaters require relief valves, which should be kept clear and never tampered with; furnace chimney and flue connecting pipe sound and tightly connected; if signs of separation are seen, have service man check the installation and make correction; erratic function of gas-fired furnace should be reported and checked out promptly.

After cleaning paint brushes with a volatile solvent, follow by washing with soap and water. Never leave the brushes suspended in an open can of the solvent inside the house.

Household cleaners are not usually flammable, but they may be poisonous, highly toxic, or cause serious burns when in contact with the skin. Oven cleaners often contain lye or similar caustic. This is true also of drain cleaners, which may also contain acids. Use rubber gloves when handling these products and avoid inhalation. Carbon tetrachloride has been condemned as a cleaning compound, but it is still found in many homes even though it can cause fatal respiratory failure.

Aerosol Spray Cautions. Aerosol cans have added great convenience to the use of many products, but caution against breathing the spray must be emphasized. This is particularly true of the enamels and lacquers, which often are so carelessly used. Best to cover the nostrils with a damp cloth and keep the face far back when spraying with these cans. This warning applies to all other aerosol products, including hair spray, anti-perspirants, and particularly insecticides.

Benzene, kerosene, varnoline and other petroleum distillates used occasionally for cleaning purposes, have been responsible for numerous tragic fires. They should be used only in the open air, and adequate precautions taken against flame or sparks that might ignite the fluid. So innocuous a chemical as furniture polish can be as lethal as strychnine if swallowed by a child.

Excessive Use of Insecticides. The more potent and dangerous insecticides are gradually being removed from the consumer market as a result of efforts by governmental agencies, including the Consumer Protection and Environmental Health Service. However, all household insecticides still require careful handling and safe storage. They should be used sparingly and under controlled conditions — that is, with ample ventilation, and when children have been excluded from the rooms. Some homemakers are so concerned about cleanliness that they go after even a single fly in the house with a battery of insect killing "bombs" or sprays, little aware that the excessive concentration of the chemical mist can be more injurious to the health of the family than a few flies or mosquitoes.

Keep all insecticides on high-up shelves, not under the sink where they can be reached by children who may be tempted to try them out on the pet dog, or even ingest the chemical themselves.

Garden insecticides, rose sprays, and similar substances, must be used cautiously. Avoid inhalation, use only briefly each time — wear a respirator if fogging a large area. Keep these chemicals in a separate cabinet in the garage or potting shed, under lock and key.

Government Curbs Pressed. The unabated incidence of accidental poisonings among small children, estimated as high as 600,000 a year, with an average of 300 deaths annually of children under the age of 5 from swallowing toxic substances, has aroused considerable alarm in Congress and government agencies. Passage of the Poison Prevention Packaging Act, the efforts of the Office of Product Safety of the Food and Drug Administration, and participation of other government agencies in safety programs, hopefully will in time curb and even eliminate the hazard.

The prime responsibility is still with the parent, as it should be. While there is no substitute for constant vigilance, many routine measures in the home can help prevent accidental poisonings and injuries.

Medicines are one part of the danger — many everyday drugs such as aspirin are extremely potent and can have a very serious effect on a child. But all other possible liquids and tablets should be kept under constant control. This is not very difficult if a definite routine is established and a specific place provided for these products.

Safe Storage of Medicines. Medicine bottles have been improved considerably; most are now in unbreakable plastic with press-fit caps. The objective is to develop containers that would be "significantly difficult" for a child to open. While some gains have been made, it remains for technology to find the foolproof container, one that just cannot be opened by a child.

The best safeguards are (1) keep all medicines and similar items, including razor blades, in a single place, never leaving them around on night tables and in kitchen cupboards, and (2) provide an out-of-reach cabinet that is either kept locked, or has an adequate latch or hook that cannot be opened by a child. An important additional detail is that there be proper lighting at this cabinet to minimize any possibility of error when removing a medicine bottle or measuring out a dose.

20

Man-Made Environmental Dangers

Man-made environmental nuisances which adversely affect large numbers of persons often can be alleviated or blocked at the start by alert, knowledgeable residents. Certain industrial, commercial, and institutional activities insidiously encroach upon a community, often with assurances that the operations will be unoffensive, but despite these pledges the area becomes polluted with noise, smoke, fumes, dust, and odors. When carried to extreme levels, these contaminations can seriously detract from the comfort if not the life style of the residents, peril their health, destroy property values, and ruin the ecology of the neighborhood's lakes, streams and open lands.

Typical examples are the construction of a cement-mixing plant occupying a comparatively small industrial site, or the exploitation of a neighborhood sand pit. In either instance, there soon is a steady stream of huge diesel trucks roaring through the area, placing children in jeopardy, spewing forth huge clouds of exhaust fumes, defacing the natural terrain, and coating everything with a layer of cement dust.

An issue that has attracted wide attention, and much opposition, is the effort of utilities to construct nuclear power plants to meet the ever-increasing demands for electricity. Siting of these plants has been resisted by nearby residents who fear exposure to radiation or catastrophic accident, and by nature conservationists because the huge amounts of heated water emitted by the atomic plants is believed to affect the ecology of marine life.

The complexity of the problem is demonstrated by the fact that objections of equal validity apply to conventional fossil fuel plants, which are condemned as a major air polluter in most cities. One protest meeting to block construction of such a plant was told that residents near a similar plant were "afraid to breathe the air"!

This issue of nuclear power plants probably will be pursued until actual experience demonstrates that there is no emission of radiation sufficient to peril health.

Industry Invited. While most towns welcome industrial development as a means of growth, providing more employment opportunities and helping to carry the local tax burden, a truly satisfactory relationship results only when industry shows a genuine concern for community problems. Though an objectionable industrial operation is not easily uprooted or curtailed, an alert residential community can make an effective stand to preserve its character and natural resources from destructive abuses.

A case in point was the Delaware town that for 15 years had suffered the sickening odors emitted by a fertilizer processing plant. The town's population had shrunk to less than 1,200 because of the unhealthy condition by the time the legal battle that had been fought against the plant was finally won in the U.S. Supreme Court. The ruling that ordered the plant to close reminded other communities to be alert if they wish to avoid similar nuisances. The lesson was clear also for industry: it must manage to function in a place and manner not to destroy established life values.

Recourse to Courts. Individual homeowners sometimes are successful in legal actions asserting a constructive nuisance that denies them the enjoyment of their homes. Such suits may be based on actual loss resulting from cracked foundations or other damage caused by machinery vibrations, blasting, and similar activities, or the deterioration of paint and sidings from chemical emissions. Damage lawsuits of this nature have been more successful in recent years than formerly.

Sues Over Smoking. An interesting example of residential power is seen in an incident in a New Jersey community where buses would park for hours at the end of their runs in front of a group of houses, their diesel engines idling continuously, emitting obnoxious exhaust fumes and causing considerable noise. After one of the homeowners went to court to seek an injunction, the bus company conceded the merits of the complaint and consented to construct an alternate parking facility.

In a similar action, a Staten Island, New York, resident sued for damages over the air pollution and noise at his home caused by buses parked in a nearby garage with their engines running throughout the night. The bus operator, in defense, stated that the condition was necessitated by the fact that the garage was unheated and that the motors were kept running to keep them from freezing. But court action brought a promise for quick remedy of the complaint.

However, lawsuits for money damages or injunctions are costly, acrimonious, and not the most expeditious way to settle such disputes. The desirable and effective way to retain a healthy community is for an alert citizenry, through civic associations, to maintain a friendly but watchful relationship with industrial management while insisting on measures to avoid polluting the local water supply, streams and atmosphere.

Fight on Plane Noise. Excessive noise is an environmental pollutant that not only menaces health but also can disturb social relationships, personality development, and individual capabilities. One of the most widespread and persistent sources of disturbing noise is caused by planes along the flight patterns near airports where the intense screech of low-flying jets can become a nightmare to local residents. Efforts of the Federal Aviation Administration to reduce this annoyance have produced slim, if any, improvements. Various scientists are wrestling with efforts to establish "standards" of noises, trying to learn how much is "too much" and at what noise level does the average person crack up, as the term goes. Meanwhile the objectionable condition accelerates steadily with the increased dependency on air travel.

Practical and effective solutions to the airport noise problem are not yet in sight. Judging by the results so far, it is not likely that there can be much reduction in the noise level caused by planes, so attention must turn to other means of abating this nuisance. One obvious measure, which may be long and costly in the making, is a complete change in the location of huge airports. These must be sited in open areas far from home settlements, and serviced by very fast ground transit such as Japan's Tokyo-Osaka railroad which makes the run at 130 miles an hour. Thus an airport even 60 miles away can be reached in perhaps half an hour running time, which is a lot quicker than the time to reach the airports located in the city perimeter on roads clogged with traffic.

Another possibility is a turn back to railroad service for short inter-city runs, thus reducing the number of flights. This can become a reality with improved railroad service. Passengers can travel by train from city center to city center quicker than by plane, for distances of 150 miles or less, when the factors of travel to airports and take-off time are considered.

Rejuvenated railroad service is still not in the cards — in fact, passenger trains are being discontinued each year in the United States, whereas in European countries rail service has been maintained at a comfortable level.

Perhaps continued and emphatic protests by noise-troubled communities will bring about some changes. One of the organizations active in this effort is the National Organization to Insure a Sound-Controlled Environment (NOISE). Community support of this organization has been growing, with a number of Chicago-area municipalities joining the organization's court suit to have the roar of jet aircraft declared a public nuisance. At the Los Angeles airport, hundreds of homes have been purchased and will be razed—a kind of solution in reverse: if you can't quiet the jets, remove the complaining residents.

Shutting Out Traffic Noise. Airliners are not the only, nor even the most obnoxious source of noise, as anyone living near a main highway can attest. The toll of truck and auto noise is incalculable, but somehow residents have managed to withstand this noise invasion of their homes without too much adverse effect. The technological solution to the traffic din—and the inevitable dirt and dust that accompanies it—is the air conditioner. Windows may be kept closed all summer long, while the air conditioner hums its reassuring song of comfort. For the really sophisticated homeowner, double-glazed prime windows help to cut the noise to a minimum, and this benefit may be extended by leaving the storm windows on all year round—at least on those windows facing the road, with air conditioners supplying all ventilation.

Factory Non Grata. When 30 residents of a Rockland County, New York, community learned that an immense aluminum can manufacturing plant was to be built in nearby Congers, they set up an Emergency Committee to Save Our Environment. Their protests to the county zoning board, which had originally given its approval for the plant, succeeded in halting construction while court action was initiated. In the words of one of the committee members: "We are asking them to stay away from where people live. They could build on other sites, but they want this one, and close their eyes to the human factors involved."

And that is the core of the problem—the human factors in industrial, airport and power plant developments.

Blast, Thump and Chatter. Hardly a neighborhood escapes the onslaught, at one time or another, of construction teams going full blast from the first crack of dawn. In some areas where hard pan rock is blasted or drilled for foundations, the noise can pile up to psychological trauma. Regulatory restrictions on these activities should give some measure of relief to affected homeowners, but often these victims are either unaware of the ordinances that are supposed to benefit them, or too timid to make an issue of the offenses. Failure to assert your rights by filing a complaint comes to the same thing as waiving them.

In nearly all communities, blasting or rock drilling that is done near residences is restricted to certain daylight hours, usually 8 A.M. to 4 P.M. The building contractor would like to keep his expensive equipment working as many hours as possible, so he stretches the time several hours beyond the deadline and even would keep the work going round the clock, if not halted. But just a word from you, and the work will stay within bounds.

Pay for Breakage. Another important point to remember is that you have a right to claim damages for any deterioration of your house caused by the construction work. It is up to the blaster to do his job in such a way that your home is not damaged, so inspect your premises carefully for cracks in the foundation and plaster walls that result from any dynamite blasting. Your insurance may cover the damage, but be ready to make a claim directly against the contractor or builder for a cash settlement. In the same situation, if your walls are threatened, you may be justified in demanding that they be shored up before any excavation or demolition work is attempted.

You may be entitled to compensation also if part of your sidewalk is taken to widen the roadway, or if nearby construction of certain types of buildings or facilities tends to lower the value of your property. This will depend largely on local or state laws, but the matter certainly merits investigation by your attorney.

Warehouse Blocked. Not all community protests involve major manufacturing plants. One drawn-out and bitter battle was fought over a plan calling for construction of a seven-story warehouse adjacent to a community of small homes.

At a series of protest meetings, speakers representing all segments of the community denounced the plan as putting the lives, health and welfare of the residents in jeopardy because of the heavy truck traffic around the clock that the warehouse would introduce. Concern was expressed also that the immense warehouse facility would provide a congregating place for loiterers at all hours of the day and night. Committees set up in the course of the protest action to obtain zoning restrictions on the property initiated a campaign to turn the area into a park with needed recreation facilities.

A related action in a nearby community sought to prevent conversion of a former skating rink into a taxi garage. The property owner's appeal to rezone the building for commercial purposes, permitting it to be used for the taxi fleet, prompted a wave of neighborhood protest which halted the project.

A NEW SCOURGE — VANDALISM

Vandalism burst forth in the 1960's like a disease after an incubation period. It was the very worst time, compounding the great stresses which existed in that period, subverting the efforts then being made to erase inequities and solve many economic, racial and social problems. The irrational and purposeless acts that sprouted almost everywhere in the country caused incalculable damage, closed down schools, added to the heavy education budgets, disrupted transportation, desecrated churches and even cemeteries, harassed families to the extent of driving many from their homes, and in not a few instances, resulted in serious injuries and even deaths.

No single explanation can be offered to cover the wild excesses committed by undisciplined or hate-crazed youngsters — and by some adults with infantile mentalities. For a period, the chief victim was the railroads; trains were bombarded with stones thrown from overpasses. Often many windows were broken and a number of passengers injured by the shattered glass. Some railroad cars went into service with all, or most, of their windows cracked or broken.

Schools became another major target — in some areas every

window in a school would be broken out each weekend—
resulting in a tremendous additional expense at a time when
school budgets were facing a cost squeeze from inflationary
pressures. Heavy window guards did not solve the problem,
only made the schools look like jails.

Then the syndrome spread out in other directions. How
can a society explain an act in which a driver was killed by
a 30-pound rock dropped on a Los Angeles freeway from a

Don't duck, this window won't break. It is not glass but Lexan, a
plastic that will withstand even big rocks. There is no complete
protection from vandalism, however.

pedestrian overpass. The response to this vicious conduct was the passage of a bill by the state legislature for placement of screen barriers along all freeway overpasses. Obviously, this measure is a palliative, and may not effectively deal with the situation.

In one New York area, the opening of a new high school, which the community had long needed and worked hard to obtain, was long delayed by repeated acts of vandalism which included cutting of the power cables used in the construction. In a nearby area, a petition campaign was started for stricter police protection against teenage vandals and young toughs accused of breaking apartment windows, disturbing the peace, vandalizing playgrounds and other objectionable acts.

Up in arms are parishioners of the many churches that have been maliciously attacked by vandals, often destroying the sanctuary, creating havoc by upturning pews and breaking everything they can get their hands on. Not a few churches have been burned to the ground, including several synagogues in Rochester, New York.

In some apartment neighborhoods, bands of hoodlums make living conditions insufferable, just for the sake of mischief—any kind they can muster—overturning garbage cans, flinging the covers at apartment windows heedless of possible injuries from broken glass, crashing entry doors, intimidating and annoying passersby.

Police are mysteriously absent during these events even when alerted by the residents, perhaps deliberately avoiding encounters which could lead to pitched battles. The arrogant toughs evidently are aware of this and seem to be challenging the authorities.

But what happens when a District Attorney's home is vandalized? Does he grin and bear it, like everyone else? The answer is yes. Queens County, N.Y., District Attorney Tom Mackell's home was vandalized twice within a few months, the windows smashed and walls smeared with paint.

Private Homes Attacked. The young hoodlums make life miserable for some homeowners who offend them for one reason or another. Particular targets, but by no means the only ones, are blacks who move into newly integrated neighborhoods or build their own homes. Such race-motivated attacks include throwing paint bombs or garbage on the house, and in some cases wreak extensive destruction of the interiors.

What Is the Cure? Knowledgeable observers recognize that the problem grows from deeply rooted social conflicts and will not be resolved by individual actions. Corrective measures take two forms: one is to lessen the damage and possible injuries by various safeguards; the other is to educate the community to cooperate in apprehending the offenders.

The Rebound. Some effective methods have been developed that may reduce the cost and peril of vandalism. One interesting development is the use by a railroad of helicopters to patrol the line's trackage and spot rock throwers. The Long Island Railroad, which spends more than a million dollars a year just for repairing smashed windows, has launched a helicopter surveillance program, linked by radio communication with ground patrol vehicles. When rock throwers, or persons tampering with railroad equipment, are spotted from the air, the crew can quickly dispatch a patrol car to the scene.

On another front, technology has come to the rescue. The Lexan plastic sheeting developed by General Electric is being installed in school windows and on trains. The transparent sheeting cannot be shattered, and thus will end great expense and peril. The clear plastic undoubtedly will come into use in homes, churches and other places that are subject to attacks by teen gangs, and provide additional security against entry.

The important social front is also receiving greater attention from prominent and influential persons. In one instance, a Catholic bishop spoke out strongly to censure vandals who damaged a $50,000 home reportedly purchased by a racially mixed family. The Most Rev. Francis J. Mugavero said he would remind "all who are involved, either as perpetrators or as witnesses of this deplorable act, that those who seek their will outside the law inevitably find that they have created a community of lawlessness in which they themselves are denied protection."

21

Insurance Protection

However much they complain about continually rising insurance premiums, policyholders are envied mightily by those who just can't get insurance at all. Many individuals have difficulty obtaining fire policies for their homes and apartment furnishings, automobile injury, theft and collision coverage, jewelry floaters, public liability, and certain other insurance needs. Only short-term accident and baggage loss insurance is readily available, at extremely high cost.

In contrast to traditional practice where insurance brokers actively sought customers, some agents now charge a fee to place policies on the grounds that the commissions they receive are inadequate for the services performed. Even with this extra charge, don't expect a broker to assist you in processing a claim for any loss unless you are an important commercial customer. On the contrary, you will more likely be urged to "forget it" and drop the claim, or face either an increase in your premium or policy cancellation.

Quite often, insurance cancellations take place even though

the policyholder has never filed a single claim. The reason may be the company's poor experience in a particular area, its discrimination against policyholders above a certain age, an adverse credit report, or some other circumstance, often merely hearsay or untrue rumor. Sometimes a company has reached its limit on policies for a certain type of coverage and cuts off writing additional business — leaving former policyholders out in the cold.

The Homeowners Policy. Combined in a single policy, the homeowners insurance provides coverage for fire, liability, burglary and certain additional hazards. The "homeowners" policy is sometimes purchased by apartment renters to get the fire protection for furnishings and liability insurance.

The fire insurance pays up to the face amount of the policy for damage to the home or garage by fire, lightning, or explosion, only if the insured amount is equal to at least 80 per cent of the replacement value of the home at the time of the loss. This requirement is explained in detail later in this chapter. Home furnishings, clothing and other possessions also are insured against fire, generally for a fixed percentage (usually 40 percent) of the home insurance.

The policy also covers loss by theft (including damage to the property by burglars), holdup, theft away from home (but not from an unlocked car), vandalism, and similar risks. Family liability pays the medical bills and protects against damage claims by persons injured, on or off your premises, from the actions of any member of your immediate family (excepting automobile accident claims).

These coverages can be extended by additional provisions in the policy, usually at slight extra cost, to insure against other risks such as falling objects (tree limbs, chimney caps, etc.) collapse of the building, pipe freezeup, water tank rupture, rain damage, and damage done by movers or tradesmen, for example. Certain risks, such as hailstorm damage, are subject to a deductible amount.

With all the range of risks included in the homeowner policy, many are still not insurable. These are for flood, landslide, earthquake, war, nuclear radiation, sewer overflow, and similar catastrophes. The federal government has begun to offer limited insurance for some of these hazards.

The Coinsurance Clause. There is considerable misunderstanding about this "80 per cent" clause in fire policies, though this provision is quite simple and it is important that the homeowner be thoroughly familiar with its requirements. The purpose of coinsurance is not to deprive the homeowner of full recovery for any loss, but rather to see that the cost of insurance risk is spread equitably over all policyholders by inducing them to carry insurance in an amount closer to the real value of their properties.

In practice, payment for fire loss is made in the proportion that the total insurance carried bears to the real (replacement) value of the home — except that 80 per cent of value is considered to be adequate coverage. If the home is insured for half of the 80 percent value, then payment for any loss will be half the actual loss. But when the full 80 percent is carried, a total loss can be reimbursed at full replacement cost up to the amount of insurance. (In some policies there are provisions regarding depreciation deductions.)

Two examples are cited to illustrate the coinsurance principle. A house insured for $20,000, its original cost, was totally destroyed by fire. Replacement cost was estimated at $50,000, so the house should have been insured for 80 percent of that amount, or $40,000. Since the actual insurance was only half of that, the insurance payment also was half, or $10,000. But if the insurance had been of the required amount, the full $40,000 of insurance would be paid. The land, and the foundation of the building, are not affected by fire, and thus do not figure in the 80 percent clause or the loss.

In the second example, a $10,000 home was partially damaged by fire. The owner, figuring most fire losses are small, had insured for only $2,000 — this was only one quarter of the $8,000 required to meet the 80 percent requirement. Repair and replacement costs of the fire damage amounted to $1,000. The insurance payment was only $250, the one fourth ratio that the insurance carried bore to the 80 per cent level.

The burden of deciding how much insurance to carry falls

upon the homeowner. The original cost has no relevance — it's the replacement (reconstruction) cost that counts, and that is indeed difficult for the average person to determine. You can't tell by the selling price, as that not only varies considerably from time to time according to market conditions, but also is much different than construction or repair costs. The services of an expert consultant may be desirable; insurance brokers may not be in any better position than you are to fix an amount, and they are parties with conflicting interests.

Keep in mind though, that in computing the replacement value, you can exclude the value of the land and the building foundation. Also, a detached garage and other buildings may be considered as separate items in computing value. But do include improvements, and also the cost of rebuilding stone or brick walls of the house, even though you think they would survive any fire, since such costs would be part of the overall computation to see whether you have carried the 80 per cent coverage.

Every homeowner should review his fire insurance coverage annually to make certain that it is in keeping with current costs and values.

Now that you've gathered the details regarding "coinsurance" look over your homeowners policy again, this time to see if there are any exceptions regarding windstorms, hail, lightning, tornadoes, riots, explosions, freezeups, smoke and water damage, glass breakage, swimming pool, TV aerial damage, falling objects, vandalism, and all the rest. There are several levels of coverage: "Extended coverage" that goes a step farther than the basic policy, and "all risk" insurance that gives you almost ultimate protection against nearly every kind of hazard except ordinary deterioration. There may be special types of protection that you need and can arrange to get, such as an "Additional Living Expenses" rider which pays for hotel accommodations and similar extra living expenses while your home is being repaired.

Supplementary Protection. The homeowners insurance also can be broadened by obtaining extra riders to protect you against such hazards as Theft Away from Home. This includes baggage, camping equipment and such, but there's a definite limit on cash, usually $250, so depend on traveler's checks. You can also get protection against credit card fraud and bank check forgery. This rider may have some deductible amount such as the first $100 of loss, but it is still invaluable for the peace of mind it affords when you carry a stack of credit cards on which you are liable for any charges.

Don't neglect to make occasional reviews of your homeowners policy, as new situations may arise that require attention. Did you install an in-ground swimming pool this year? Make sure that your policy includes coverage: otherwise take immediate steps to get a swimming pool rider.

Attention, Boat Owners! Your homeowners policy most likely insures your boat, its motor, furnishings, and trailer, regardless of size, up to $500 against fire, theft and other perils except wind and hail. The personal liability and medical payments coverage in the policy apply only to rowboats, sailboats less than 26 feet long, outboard motorboats with up to 25 horsepower, and inboards up to 50 horsepower. (These horsepower limitations vary in some states.) You can expand this protection with an Outboard Motor and Boat policy, or a Yacht policy, providing "all risk" protection from a wide variety of marine hazards.

"Floaters." Since the theft protection under the homeowners policy excludes or puts narrow limits on such valuables as jewelry, furs, cameras, and treasures like coin and stamp collections, a separate policy is issued which lists and identifies each item that is covered. The "floater" policy does not have the usual limitations as to the place or circumstances of loss; that is, a loss is recognized even if it results from forgetfulness such as leaving a ring on a lavatory sink in a theater, or neglecting to check a fur wrap at a restaurant. Premium rates for floater policies are rather high, and even so, these policies are difficult for some persons to obtain.

Art collections, rare violins, irreplaceable antique furniture can be insured by special arrangement which will take into account the location of the items, the extent and nature of security arrangements, the frequency of shipments, and other factors. Lloyds of London is famous for writing these special individual policies but at high premium rates.

Family Liability Insurance. This covers claims for injuries and property damage that occur either at your home or elsewhere as a result of the actions by any member of your immediate family. Thus coverage includes injuries suffered by guests or other visitors at your home (the postman, a domestic worker, or a canvasser, for example) as a result of some condition such as a loose stair rail, poor lighting, a projecting casement window, snow or ice sliding off the roof, etc.). Unless specifically excluded, the liability insurance also covers hazards of the swimming pool, and bites inflicted by the family's pet dog.

Claims for injuries caused while playing golf, or surfboarding, and other activities of family members, are covered. But injuries inflicted deliberately, such as in a fistfight or similar entanglement, are not.

The insurance company pays medical expenses, within specified limits, without regard to who is at fault, and also is obligated to provide legal defense against claims. A single act of carelessness can result in extremely high damage awards, so the important thing is to carry an adequate amount of liability insurance to protect your home and financial security.

Auto Insurance. "Get a horse!" was the derisive advice to owners struggling with the temperamental early automobiles. Now the rubber-tired wheel has turned full circle and maybe buyers in the near future will again go shopping for a horse and buggy—or maybe a bicycle—instead of a car.

If that happens—and it's not so unlikely—the reason may lie more with insurance difficulties than with any other single factor. The soaring premium rates, the arbitrary rejection of applicants or cancellation of policies, and the financial pressures on the companies themselves, all mount up to the equivalent of a state of siege, with no relief in sight.

Insurance companies can refuse to sell a policy to any individual without explanation. Poor driving record, a history of traffic violations or criminal charges, unstable employment, slow payment of bills or excessive debts, personal bankruptcy, poor moral conduct—even just mature age—any of these may be cause for rejection. Some companies simply refuse to cover sports cars or mini cars. Then the driver must seek insurance from the "Assigned Risk Pool." He is automatically labeled a "bad risk" and his insurance dropped at the first opportunity.

Before applying for "Assigned Risk" coverage, a rejected applicant should make a determined canvass of other brokers and companies with the hope of finding one with less bias toward his situation. An effective measure is to apply for both homeowner and auto policy together, making the deal more attractive for the broker's attention. Some companies also are more liberal in their judgment of individuals than are others, or receptive to a certain amount of business from a different area for diversification purposes.

But there also are some companies that *guarantee* renewal of auto insurance coverages. Allstate Insurance Company offers a renewal guarantee for five years after the policy has been in force for 60 days. This guarantee is not available in certain states (Texas, North Carolina, Louisiana, Virginia, Massachusetts) or to commercially rated vehicles and in most states to assigned risk policyholders. The Allstate policy also provides automatic coverage for policy holders while driving rented cars or cars borrowed occasionally from friends.

No Fault. Supporters of an innovative concept in handling auto accident claims, the "No Fault" system, hope for dramatic reductions in court litigation, insurance costs, and premium rates. Initially opposed by various groups, including trial attorneys who contend that the system does not provide fair compensation for the injured or punish the persons responsible for negligence, the no fault concept is widely regarded as the best means in sight to correct current abuses and solve many pressing problems.

Car Pools. If you get to work as a member of a car pool, or your children are driven to school by a mothers' car pool, the insurance details need to be reviewed. Any car used for this purpose should have at least $100,000 and $300,000 liability with $25,000 medical payments coverage. However, a driver who accepts reimbursement for gas, tolls, and other expenses, is heading for trouble—the riders in that case may be considered as "paying passengers" rather than as guests, and thus have a much broader basis for claims in the event of accidental injuries. Any such car pool arrangements should be checked with an insurance consultant for interpretation to avoid misunderstanding and devastating lawsuits.

Good advice is given to mothers regarding school car pools by safety advocate Annette Shelness who says, "We take too much for granted when we assume that every car pool mother is automatically a good driver and that her vehicle is well maintained in safe condition." Mrs. Shelness recommends that car pool members reach agreement on the minimum insurance each driver should have, the safety standards applied to the cars, the pickup procedures, and the children's behavior en route. Pool cars should have periodic checkups at approved garages; discipline rules for pupils strictly enforced and supported by the parents, and pickup stops selected that are clear of traffic hazards.

Additionally, recommendations include close observation of each mother's driving capability. Does she keep her cool under the stress of half a dozen or more yelping children? Does she respond promptly to changing traffic conditions? Does she comply carefully with traffic rules, or does she try to "beat the lights"? And very important, is there any sign of a drinking problem or the effects of a medication?

Auto Liability Insurance. Compulsory in many states, such policies have been called upon to pay lawsuit judgments of astronomical levels. A single accident may result in a judgment far above the amount of insurance, often to the extent of wiping out the car owner's assets, including his home. (The personal liability clause in the homeowner policy does not cover automobile accidents.)

For families with a responsible financial history (no bankruptcies, judgments, or other black marks) extra high liability coverage can be obtained at nominal additional charge. The protection may be for $100,000 and $300,000, (that is, $100,000 for an injury to a single person, $300,000 for injuries to two or more persons per occurrence), and may even go as high as $500,000 and $1,000,000, and is well worth the extra cost.

An additional benefit in carrying the extra high liability coverage is that the heavily committed insurance company will stick with you to defend any lawsuit to the bitter end, rather than settling with the plaintiff for a modest insured amount, and leave you facing additional damages on your own.

Property damage coverage, which is part of the basic liability insurance, also is moving into ever-higher expense levels. The most minor accident can result in repair bills going into the hundreds of dollars. Tests have shown that a car bumped in the rear at a speed of only five miles an hour, can suffer damage amounting to over $200.

Collision Insurance – Is It Worth the Cost? An investment of some $3,000 to $5,000 in a car certainly warrants collision insurance, which will pay for repairs to your own car if damaged in an accident. A sideswipe accident that crumples the doors and fenders can bring a repair bill almost as large as the original cost of the car. But the collision insurance should be regarded only as protection against major damage, not for payment of average small repairs. A $100 deductible lowers the insurance premium considerably and provides the needed protection against a huge damage expense.

Theft and Comprehensive. This clause covers theft of the car, fire, and other non-collision loss such as cracked windshield and malicious damage. Theft of accessories is covered when there are signs of forcible entry. The insurance also reimburses for damage caused by the thieves, such as dented trunk lids, damaged ignition locks and coils, and forced windows.

Product Sources

CODE TO PRODUCT SOURCES

1. Burglar alarms, systems and complete units.
2. Burglar alarm components, accessories, and installation tools.
3. Personal defense products.
4. Locks.
5. Door and window latches, chains, braces, interviewers.
6. Fire alarm systems.
7. Fire protection sprinklers.
8. Auto locks and alarms.
9. Safes, strongboxes.
10. Window guards, gratings, escape ladders.

CODE

9 Accountants Supply House
528 Rockaway Ave.,
Valley Stream, N.Y. 11582

1 Adco Alarms
150 West 28 St., New York, N.Y. 10001

1 Advanced Devices Laboratory, Inc.
701 Kings Row, San Jose, Cal. 95112

5 Ajax Hardware Corp.
825 S. Ajax Ave., City of Industry, Cal. 91747

2 Alarm Components Distributors Co.
33 New Haven Ave., Milford, Conn. 06460

1,2 Alarm Device Mfg. Co. (Ademco)
165 Eileen Way, Syosset, N.Y. 11791
Alarm Services Co.
5228 Continental Dr., Rockville, Md. 20853

1 Alarmtronics Engineering, Inc.
154 California St., Newton, Mass. 02195
American District Telegraph Co. (ADT)
155 Avenue of the Americas,
New York, N.Y. 10013

10   American Security Products Co.
    16720 S. Garfield Ave., Paramount, Cal. 90723
1   Applied Electro-Mechanics, Inc.
    2350 Duke St., Alexandria, Va. 22314
    Arrow Fastner Co.
    Mayhill, Saddle, N.J. 07663
2   Arrowhead Enterprises, Inc.
    P.O. Box 191, Bethel, Conn. 06801
1   A.T.A. Control Systems, Inc.
    340 W. 78 Rd., Hialeah, Fla. 33014

1   Ballistics Control Corp.
    39-50 Crescent St.,
    Long Island City, N.Y. 11101
10   Julius Blum & Co.
    Carlstadt, N.J. 07072
1   Bourns Security Systems, Inc.
    681 Old Willets Path, Smithtown, N.Y. 11787
9   John D. Brush Co.
    900 Linden Ave., Rochester N.Y. 14625
5   Builders Brass Works
    P.O. Box 2043, Los Angeles, Cal. 90054

1,2   Chicago Fire & Burglar Detection, Inc.
    646 Roosevelt Ave., Glen Ellyn, Ill. 66137
1   Continental Instruments Corp.
    3327 Royal Ave., Oceanside, N.Y. 11572
1   Control Dynamics, Inc.
    810 W. Collins Ave., Orange, Cal. 92667
2   S.H. Couch, Inc.
    3 Arlington St., N. Quincy, Mass. 02171

2   Davis Aircraft Products, Inc.
    Northport, N.Y. 11768
1   Delphi Corp.
    1701 N. Fort Myer Drive, Arlington, Va. 22209
2   Delta Products
    P.O. Box 1147, Grand Junction, Col. 81501

1,2   Design Controls, Inc.
      75 Sealey Ave., Hempstead, N.Y. 11550
  1   Devenco, Inc.
      150 Broadway, New York, N.Y. 10038
1,2   Dialalarm, Inc.
      7315 Lankersheim Blvd.,
      N. Hollywood, Cal. 91609
  1   Dictagraph Security Systems, Inc.
      733 Mountain Ave., Springfield, N.J. 07081
  9   Diebold, Inc.
      Canton, Ohio 44702
 10   Dilworth Mfg. Co.
      Box 285, West Chester, Pa.
  1   Dimension, Inc.
      P.O. Box 109, Chantilly, Va.
  2   Diversified Mfg. Co.
      P.O. Box 2202, Burlington, N.C. 27215

  4   Eagle Lock Co.
      Terryville, Conn.
  6   The Edwards Company
      Connecticut Ave., Norwalk, Conn. 06856
  1   Electronic Enterprises, Inc.
      65 Seventh Ave., Newark, N.J. 07104
1,2   Electronic Locator Co.
      350 Gotham Pkway, Carlstadt, N.J. 07072
  1   Emerson Electric Company
      8100 Florissant, St. Louis, Mo. 63136
  6   EverGard Fire Alarm Co.
      601 Susquehanna Ave., Philadelphia, Pa. 19122

  6   Falcon Alarm Co.
      1137 Route 22, Mountainside, N.J. 07092
  3   J.F. Farmer Mfg. Co.
      P.O. Box 18024, Indianapolis, Ind. 46218
  6   Fenwal, Inc.
      Ashland, Mass. 01721
  6   Fire Alarm Thermostat Co.
      333 Lincoln St., Hingham, Mass. 02043

6 Fire Alert Co.
505 W. 40 Ave., Denver, Col. 80216
6 Fire Control Co.
703 Thornton St., Wilmington, Del. 80216
6 Fire-Lite Alarms, Inc.
190 Fulton Terrace, New Haven, Conn. 06504
6 Firemark, Inc.
9100 W. Belmont Ave., Franklin Park, Ill. 60131
10 Flair Products
2409 Main St., Evanston, Ill.
4 Fox Police Lock Co.
46 West 21 Street, New York, N.Y. 10010
6 F.S.I. Security Systems, Inc.
223 Arch St., Philadelphia, Pa. 19106

6 Gardsman Corp.
160 Fifth Ave., New York, N.Y. 10010
10 General Electric Co.
1 Plastics Ave., Pittsfield, Mass. 01201
Distributors of General Electric's Lexan:
Pralcoa Glass Co.
2303 Second Ave., New York, N.Y. 10023
Cadillac Plastic & Chemical Co.
2305 W. Beverly Rd.,
Los Angeles, Cal. 90057
313 Corey Way, South,
San Francisco, Cal. 94080
823 Windsor St., Hartford, Conn. 06108
Binswanger Glass Co.
1201 N. 20 St., Birmingham, Ala. 35204
3 Gray Mfg. Co.
Tecumseh, Mich. 49286

Hammacher Schlemmer Co.
145 E. 57th St. New York, N.Y. 10022
Heath Co.
Benton Harbor, Mich. 49022
5,10 J.W. Holst, Inc.
1005 East Bay St., East Towas, Mich. 48730

1   Holmes Electric Protection Co.
    370 Seventh Ave., New York, N.Y. 10001
5   Home Protector Mfg. Co.
    P.O. Box 425, Rico Rivera, Cal. 90661
1,6   Honeywell
    2701 Fourth Ave.S., Minneapolis, Minn. 55408
5,10   Hotchkiss Products, Inc.
    New Harbor, Main 04554
10   R.H. Hutchinson & Co.
    2610 Sylvan Ave., Dallas, Texas 75212

2   Ideal Industries, Inc.
    Sycamore, Ill. 60178
5   Ideal Security Hardware Co.
    215 East 9 St., St. Paul, Minn. 55101
10   IKG Industries
    539 10th St., Kansas City, Mo. 66105
4   Illinois Lock Co.
    302 W. Hintz Rd., Wheeling, Ill. 60090
1   IMC Microwave Corp.
    33 River Rd., Cos Cob, Conn. 06907
2   Informer Alarm Supplies Co.
    200 S. Eden St., Baltimore, Md. 21231
1   Interstate Engineering Co.
    522 E. Vermont Ave., Anaheim, Cal. 92805
4   Intertrade Industries, Ltd. (Abloy Locks)
    5000 Buchan St., Montreal, Quebec 9, Canada

10   Kentucky Metal Products Co.
    Preston St., Louisville, Ky. 40213
6,7   Walter Kidde Co.
    Bellville, N.J. 07109
4   M.D.Kramer Locksmith Supplies
    535 Liberty Ave., Brooklyn, N.Y. 11207
4   Kwikset Sales & Service
    516 E. Santa Ana St., Anaheim, Cal. 92803

2,3,5   LB Distributors
    P.O. Box 15325, San Francisco, Cal. 94115

1,2,6 Lee Electric Co.
309 51 Street, West New York, N.J. 07093
5 Loxem Mfg. Co.
Midland Park, N.J. 07432
5 Luster Line Products Co.
Richmond & Morris Sts., Philadelphia, Pa. 19125

1,3,5 Macy's Herald Square
New York, N.Y. 10001
Marvin-Hall Co.
9 Crowby Street, New York, N.Y. 10013
9 Meilink Steel Safe Co.
P.O. Box 2567, Toledo, Ohio 43606
4 Miracle Lock Corp.
200 Shames Dr., Westbury, N.Y. 11590
Mosler Safe Co.
40 W. 40th Street, New York, N.Y. 10018
1 Motorola Instrumentation & Control Co.
P.O. Box 5409, Phoenix, Ariz. 85010
1 MRL Inc.
7227 Lee Highway, Falls Church, Va. 22046

1 National Alarm Products
1166 Hamburg Tpk., Wayne, N.J. 07470
6 National Fire Protection Association
60 Batterymarch St., Boston, Mass. 02110
1 National Guard Products Co.
540 North Pkway, Memphis, Tenn. 38107
National Safety Council
425 N. Michigan Ave., Chicago, Ill. 60611
1 Nebetco Engineering
1107 Chandler Ave., Roselle, N.J. 07203
1,2,6 New England Home Security Co.
Naugatuck, Conn.
9 Nor-Gee Corp.
Box 6, Lancaster, N.Y. 14086
2 Notifier Corp.
3700 N. 56 St., Lincoln, Neb. 68501

5 Novel Products
31 Second Ave., New York, N.Y. 10003
1,6 NuTone Inc.
Cincinnati, Ohio 45227

4 Preso-matic Lock Co.
8228 W. 47 St., Lyons, Ill. 60534
4 Proof Lock International
555 Madison Ave., New York, N.Y. 10022
2 Protecto Alarm Sales
P.O. Box 357, Birch Run, Mich. 48415
Puretec, Inc.
2200 Centinele Ave., Chicago, Ill. 60616
6 Pyrotecto, Inc.
349 Lincoln St., Hingham, Mass. 02043
6 Pyrotronics, Inc.
2343 Morris Ave., Union, N.J. 07083

2 Red Comet, Inc.
P.O. Box 272, Littleton, Col. 80120
7 Reliable Automatic Sprinkler Co.
525 N. MacQuesten Pkway.,
Mount Vernon, N.Y. 10552
10 Rose Mfg. Co.
2700 W. Barberry Place, Denver, Col. 80204
4 Russwin Locks
New Britain, Conn. 06050

2 Safeway Products
92 Brighton 11th St., Brooklyn, N.Y. 11235
Safe Hardware Corp.
102 Washington St., New Britain, Conn. 06051
4 Sargent & Greenleaf, Inc.
24 Seneca Ave., Rochester, N.Y. 14621
1 Scientific Security Systems, Inc.
455 Woodland Hills Bldg., Jackson, Miss. 39216
1,2,4,5,6,9 Sears Roebuck
Philadelphia, Pa. 19132

1,2,3,4,5   Security Hardware Service Co.
          1322 Haight St., San Francisco, Cal. 94117
     8   Sensitronics Corp.
          Brooklyn, N.Y.
  3,10   Spartan Sales Co.
          945 Yonkers Ave., Yonkers, N.Y. 10704
  1,3,5   The Stanley Works
          New Britain, Conn. 06050
     1   Sunset Equipment Co.
          90 Magnolia Ave., Westbury, N.Y. 11590
     1   Systron-Donner
          6767 Dublin Blvd., Dublin, Cal. 94566

     4   Taylor Lock Co.
          2034 W. Lippincott St., Philadelphia, Pa. 19132
     2   Theft Prevention Co.
          Cupertino, Cal. 95014
   1,4   3M Co.
          3M Center, St. Paul, Minn. 55101

     4   U-Change Lock Systems
          5505 N. Brookline,
          Oklahoma City, Okla. 73112
     1   United States Research Co.
          1920 L St. N.W., Washington, D.C. 20036
     1   Universal Security Instruments, Inc.
          1315 Pratt St., Baltimore, Md. 21231

          Wessel Hardware Corp.
          Erie Ave. at D St., Philadelphia, Pa. 19134
          J.C. Whitney & Co.
          1917 Archer Ave., Chicago, Ill. 60616
     4   Weiser Company
          4100 Ardmore Ave., South Gate, Cal. 90280
          Windsor Security Systems, Inc.
          1695 Bonne Ave., Bronx, N.Y. 10460

Booklets and Product Literature

Home Fire Safety Survey. An evaluation of the hazard locations in a home, for the purpose of most effectively locating automatic fire alarms and protective devices. Falcon Alarm Company, 1137 Route 22, Mountainside, N.J. 07092.

Security Planning for Buildings. A guide for utilizing security systems and devices in commercial and industrial buildings. Booklet No. 54-0349. Honeywell Company, Inquiry Supervisor, 2701 Fourth Ave. S., Minneapolis, Minn. 55408. *Fire Detection Systems,* Honeywell Booklet No. 54-0362, explains various approaches to fire protection planning, and compares automatic systems.

Lafayette Radio & Electronics' free 468-page catalogue of electronic components, burglar, fire, and auto alarms, wire, tools, and other equipment. The catalogue is a practical encyclopedia of up-to-date electronic items. Write to Lafayette, 111 Jericho Tpk., Syosset, N.Y. 11791.

Allied Radio, 100 N. Western Ave., Chicago, Ill. 60680, will send you free its current catalogue of over 500 pages describing many of the items needed for alarm installations and other home improvements, including, tools, wire, and complete alarm units.

The Security Rated Home. Attractive booklet in color containing suggestions for improving home protection, and for securing your valuables. From Schlage Lock Company, P.O. Box 3324, San Francisco, Cal. 94119. A very useful booklet, also available from Schlage Lock Company, is the *Residential Security Manual,* listing the specific types of locks that should be placed on various doors, including all entrance doors, interior doors, garage doors, and sliding glass doors.

Staple tacking isn't especially a fine art, but there's a right way and a wrong way to fasten those thin circuit wires. A useful instruction booklet on the use of stapling guns is offered by Swingline, Inc., 32-00 Skillman Avenue, Long Island City, N.Y. 11101.

The manufacturer of the famous T-18 wire tacker is Arrow Fastener Company, 271 Mayhill Street, Saddle Brook, N.J. 07663. Write for an interesting catalogue containing important information on using this stapling gun. Information is supplied on how to clear the staple passages when they become clogged.

Household Fire Warning Systems, issued by the National Fire Protection Association. Describes types of fire warning installations, lists manufacturers of fire detectors whose products are suitable for home fire alarm use, including effective distance specifications. The booklet also contains sections on the standards for installation, maintenance, and use of a household fire warning system in various circumstances. Price 50 cents, from National Fire Protection Association, 60 Batterymarch St., Boston, Mass. 02110. Request also a copy of the catalogue of *Pamphlet Editions of the Fire Codes Standards,* listing publications that may cover your specific circumstances and problems.

Power Tool Safety Rules, explaining the proper handling and servicing of various power tools used around the home.

Available from Outdoor Power Equipment Institute, 700 Marcellus St., Syracuse, N.Y. 13201.

Booklet titled *The House Is On Fire* and one called *Emergency: A Pocket Guide to Emergency Action.* Available from the National Safety Council, 425 N. Michigan Ave., Chicago, Ill. 60611.

Landscaping for Fire Protection. Information for owners of homes in forested areas. From Agriculture Extension Service, University of California, Berkeley, Cal. 94720.

The Greatest Storm On Earth . . . HURRICANE, issued by Environmental Science Services Administration, (ESSA) Department of Interior, Washington, D.C. 20005.

A catalogue that is as interesting and instructive as any manual is *Wood Boring Tools* issued by The Irwin Auger Bit Co., of Wilmington, Ohio. The catalogue shows a wide range of drill bits, door lock bits, augers for almost every purpose, including the very long electrician bits that will be very helpful in the burglar alarm installation.

Index